GCSE Success

REVISION GUIDE

Maths Foundation Tier

Fiona Mapp

Contents

Number

Revised

- 4 Types of numbers
- 6 Positive & negative numbers
- 8 Working with numbers
- 10 Fractions
- 12 Decimals
- 14 Rounding
- 16 Percentages 1
- 18 Percentages 2
- 20 Fractions, decimals & percentages
- 21 Using a calculator
- 22 Approximations & checking calculations
- 24 Ratio
- 26 Indices
- 28 Practice questions

Algebra

Revised

- 30 Algebra
- 32 Equations 1
- 34 Equations 2
- 36 Patterns, sequences & inequalities
- 38 Formulae
- 40 Straight-line graphs
- 42 Curved graphs
- 44 Interpreting graphs
- 46 Practice questions

Geometry and measures

Revised

- 48 Shapes
- 50 Solids
- 52 Constructions
- 54 Symmetry
- 55 Loci & coordinates in 3D
- 56 Angles
- 58 Bearings & scale drawings
- 60 Transformations 1
- 62 Transformations 2
- 64 Measures & measurement 1
- 66 Measures & measurement 2
- 68 Pythagoras' theorem
- 70 Area of 2D shapes
- 72 Volume of 3D shapes
- 74 Practice questions

Statistics and probability

Revised

- 76 Collecting data
- 78 Representing data 1
- 80 Representing data 2
- 82 Scatter graphs & correlation
- 84 Averages 1
- 86 Averages 2
- 88 Probability 1
- 90 Probability 2
- 92 Practice questions

- 94 Answers
- IBC Index

Types of numbers

Squares and cubes

Square numbers

Any number raised to the **power 2** gives a **square number**. For example, $6^2 = 6 \times 6 = 36$ (six squared).

Square numbers include:

1	4	9	16	25	36	49	64
(1×1)	(2×2)	(3×3)	(4×4)	(5×5)	(6×6)	(7×7)	(8×8)

81	100	121	144	169	196	225
(9×9)	(10×10)	(11×11)	(12×12)	(13×13)	(14×14)	(15×15)

Square numbers can be illustrated by drawing squares:

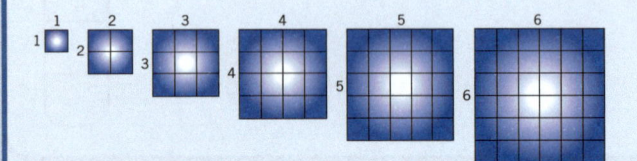

You need to know square numbers up to 15^2.

Cube numbers

Any number raised to the **power 3** gives a **cube number**. For example, $5^3 = 5 \times 5 \times 5 = 125$ (five cubed).

Cube numbers include:

1	8	27	64	125	1000
(1×1×1)	(2×2×2)	(3×3×3)	(4×4×4)	(5×5×5)	(10×10×10)

Cube numbers can be illustrated by drawing cubes:

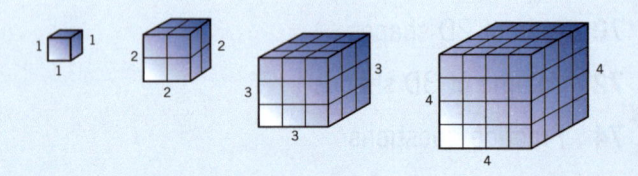

It is important that you recognise square and cube numbers because they often appear in number sequence questions.

Square roots and cube roots

$\sqrt{}$ is the **square root** sign. Taking the square root is the opposite of squaring.

For example, $\sqrt{25} = \pm 5$ since $5^2 = 25$, or $(-5)^2 = 25$

$\sqrt[3]{}$ is the **cube root** sign. Taking the cube root is the opposite of cubing.

For example, $\sqrt[3]{8} = 2$ since $2^3 = 8$

Multiples

Multiples are the numbers that appear in multiplication tables.

For example, multiples of 5 are
5, 10, 15, 20, 25, …

Multiples of 8 are
8, 16, 24, 32, 40, …

Reciprocals

The **reciprocal** of a number $\frac{a}{x}$ is $\frac{x}{a}$ ($= x \div a$). Multiplying a number by its reciprocal always gives 1. Zero has no reciprocal, because division by zero is not defined.

For example:
- The reciprocal of $\frac{2}{3}$ is $\frac{3}{2}$
- The reciprocal of 4 is $\frac{1}{4}$ (4 is the same as $\frac{4}{1}$)
- To find the reciprocal of $1\frac{2}{3}$, first put it in the form $\frac{a}{x}$ ($1\frac{2}{3} = \frac{5}{3}$), then invert it to give $\frac{3}{5}$

Factors and prime numbers

Factors
Factors are whole numbers that **divide exactly** into another number. For example, the factors of 20 are 1, 2, 4, 5, 10, 20. To find all the factors of a number, start at 1 and divide by each whole number in turn. Factors can be split up into factor pairs. For example, for the factors of 20:

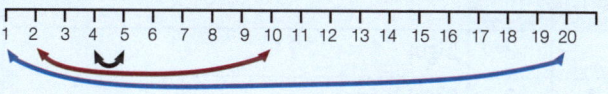

1 2 3 4 5 6 7 8 9 10 11 12 13 14 15 16 17 18 19 20

So, 1 × 20 = 20   2 × 10 = 20   4 × 5 = 20

Prime numbers
A **prime number** is a number that has only two factors, **1 and itself**. Note that 1 is not a prime number.

The prime numbers up to 20 are 2, 3, 5, 7, 11, 13, 17 and 19.

Prime factors

Prime factors are factors that are prime. All whole numbers, except prime numbers, can be written as products of their prime factors. For example, the diagram below shows the prime factors of 360:

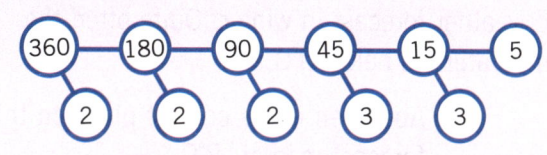

360 — 180 — 90 — 45 — 15 — 5
  2   2   2   3   3

❶ Divide 360 by its first prime factor, 2.
❷ Divide 180 by its first prime factor, 2.
❸ Keep on going until the final number is prime.
❹ As a product of its prime factors, 360 can be written as:
2 × 2 × 2 × 3 × 3 × 5 = 360
or in **index notation** (using powers):
$2^3 × 3^2 × 5 = 360$

Highest common factor (HCF)
The **largest factor** that two numbers have in common is called the **HCF**.

Example
Find the HCF of 84 and 360.

❶ Write the numbers as products of their prime factors and ring the common factors.
84 = 2 × 2 ×   3   × 7
360 = 2 × 2 × 2 × 3 × 3 × 5
❷ These give the HCF = 2 × 2 × 3 = 12

Lowest common multiple (LCM)
The **LCM** is the lowest number that is a multiple of two or more numbers.

Example
Find the LCM of 6 and 8.

❶ Write the numbers as products of their prime factors.
8 = 2 × 2 × 2
6 = 2   × 3
8 and 6 have a common prime factor of 2. It is only counted once.
❷ You take one value from each column, so the LCM of 6 and 8 is 2 × 2 × 2 × 3 = 24

Quick test

❶ List the prime numbers between 10 and 30.
❷ Find the HCF and LCM of 24 and 60.
❸ Find a) $\sqrt{64}$ b) $\sqrt[3]{216}$
❹ Write down the reciprocals of the following:
a) $\frac{3}{4}$ b) $\frac{x}{p}$ c) 5   d) $\frac{1}{10}$

Positive & negative numbers

Place value in whole numbers

The value of a digit in a whole number depends on its position in the number. This is called its **place value**.

Example
Gill buys some premium bonds for £1050. Write this figure in words.

Place value of digits				Number
1000 Thousands	100 Hundreds	10 Tens	1 Units	
			8	Eight
		7	9	Seventy-nine
	3	4	5	Three hundred and forty-five
6	3	2	9	Six thousand three hundred and twenty-nine

This number is one thousand and fifty pounds.

Integers

Integers are a set of whole numbers {..., -3, -2, -1, 0, 1, 2, 3,...}. They include **positive numbers**, **negative numbers** and zero.

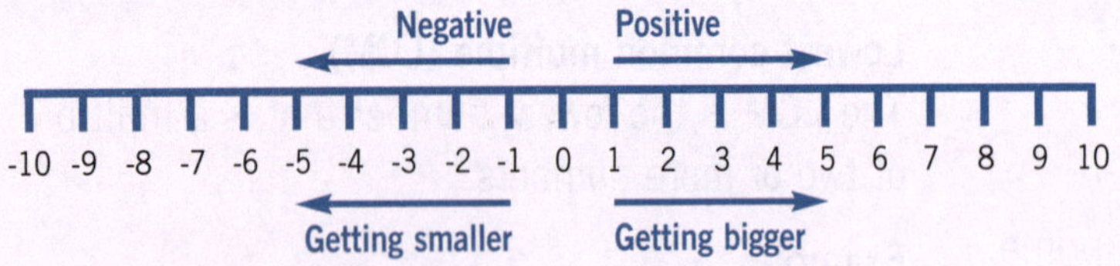

```
       Negative          Positive
←─────────────────┼─────────────────→
-10 -9 -8 -7 -6 -5 -4 -3 -2 -1  0  1  2  3  4  5  6  7  8  9  10
←─────────────────   ─────────────────→
    Getting smaller     Getting bigger
```

Positive numbers are above zero; negative numbers are below zero.

For example:
-10 is smaller than -8 or -10 < -8
-4 is bigger than -8   or -4 > -8
2 is bigger than -6   or 2 > -6

Positive and negative numbers are often seen on the weather forecast in winter. Quite often the temperature is below 0°C.

Aberdeen is the coldest place on this forecast map at -8°C. London is 6 degrees warmer than Manchester.

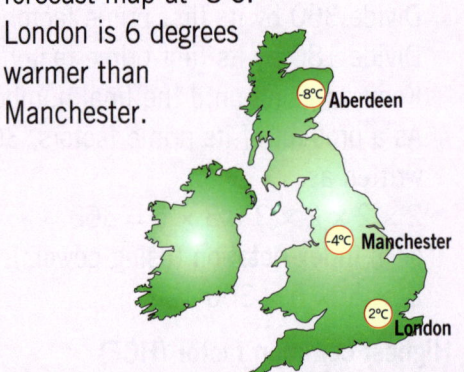

-8°C Aberdeen
-4°C Manchester
2°C London

Multiplying and dividing positive and negative numbers

Multiply and divide positive and negative numbers as you normally would. Then find the sign for the answer using these rules:
- Two **like** signs (both + or both -) give a positive answer.
- Two **unlike** signs (one + and the other -) give a negative answer.

For example:
-6 × (+4) = -24   -12 ÷ (-3) = 4
-6 × (-3) = 18   20 ÷ (-4) = -5

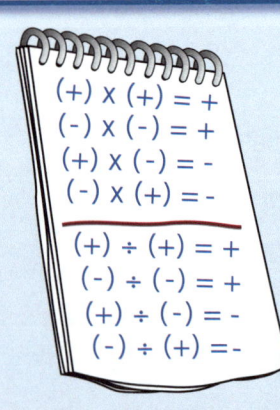

(+) × (+) = +
(-) × (-) = +
(+) × (-) = -
(-) × (+) = -

(+) ÷ (+) = +
(-) ÷ (-) = +
(+) ÷ (-) = -
(-) ÷ (+) = -

You need to remember the rules of multiplication and division. You will find these laws useful when multiplying out brackets in algebra.

Adding and subtracting positive and negative numbers

Number lines can help you to understand the concept of adding and subtracting positive and negative numbers.

For example, the temperature at 6am was -5°C. By 10am it had risen 8 degrees.

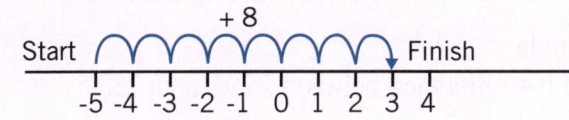

Start +8 Finish
-5 -4 -3 -2 -1 0 1 2 3 4

So the new temperature was 3°C.

Example
Find the value of -2 – 4

This represents the **sign** of the number. Start at -2.

-2 – 4

This represents the operation of **subtraction**. Move 4 places to the left.

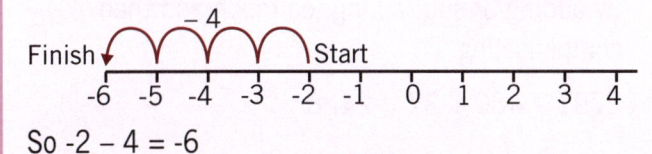

Finish −4 Start
-6 -5 -4 -3 -2 -1 0 1 2 3 4

So -2 – 4 = -6

If you find working with negative numbers difficult, sketch a quick number line to help you.

When the number to be added (or subtracted) is **negative**, the normal direction of movement is **reversed**. For example:

-4 – (-3) is the same as -4 + 3 = -1

The negative changes the **direction**.   Move 3 places to the **right**.

When two + or two – signs are together in an equation, these rules are used:

+ (+) − +
− (-) − + } **like** signs give an **addition**

+ (-) − −
− (+) − − } **unlike** signs give a **subtraction**

For example:
-6 + (-2) = -6 – 2 = -8
4 – (-3) = 4 + 3 = 7
-2 – (+6) = -2 – 6 = -8
9 + (-3) = 9 – 3 = 6

Negative numbers on a calculator

The +/− or (-) key on a calculator gives a negative number. For example, to get -6, press

 6   +/− or (-)   6

Make sure you know how to enter negative numbers in your calculator.

Look at this calculation:
-4 – (-2) = -2

This is keyed in the calculator as follows:

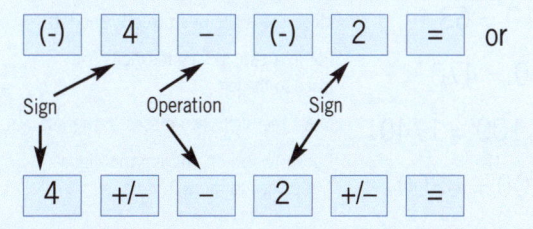

 (-)   4   −   (-)   2   =   or

Sign   Operation   Sign

 4   +/−   −   2   +/−   =

Quick test

1. If the temperature was -12°C at 2am, and it increased by 15 degrees by 11am, what was the temperature at 11am?

2. Work out the following, without using a calculator.
   a) -2 – (-6)
   b) -9 + (-7)
   c) -2 × 6
   d) -9 + (-3)
   e) -20 ÷ (-4)
   f) -18 ÷ (-3)
   g) 4 – (-3)
   h) -7 + (-3)
   i) -9 × -4

Working with numbers

Addition and subtraction

When you add and subtract integers, line up the place values one on top of the other.

Example
Work out 5263 + 398.

```
  5263
   398 +
  ----
  5661
   1 1
```
Line up the numbers first. Add the units, then the 10s, etc.

The 1 is carried here into the tens column.

This addition can be checked mentally by using **partitioning**. An empty number line can help.

5263 + 400 − 2 = 5661

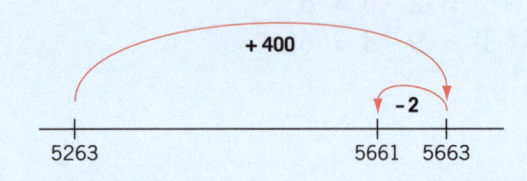

Subtracting is also known as **finding the difference**.

Example
Find the difference between 2791 and 363.

```
  2⁸7⁹¹
   363 −
  ----
  2428
```

In the units column, subtracting 3 from 1 won't work, so borrow 10 from the next column. The 9 becomes an 8 and the 1 becomes 11.

Compensation can be used to check the answer, by adding or subtracting too much and then compensating.

2791 − 400 + 37 = 2428

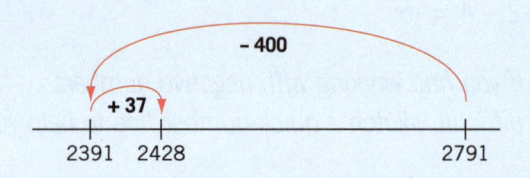

Multiplication and division by 10, 100, 1000

To **multiply** by 10, 100, 1000, etc. move the digits one, two, three, etc. places to the left and put in zeros if necessary. For example:

6.3 × 10 = 63 — Move the digits one place to the left.
47 × 10 = 470 — Put in a zero, moving the digits one place to the left.
17.4 × 100 = 1740 — Move the digits two places to the left.
68 × 100 = 6800 — Put in two zeros, moving the digits two places to the left.

When multiplying by multiples of 10 (e.g. 20, 30, 700), the same rules apply, except you multiply the numbers first, then move the digits to the left. For example:

60 × 30
= 60 × 3 × 10
= 180 × 10 = 1800

or, 60 × 30
= 6 × 10 × 3 × 10
= 18 × 100 = 1800

5.2 × 20
= 5.2 × 2 × 10
= 10.4 × 10 = 104

To **divide** by 10, 100, 1000, etc. move the digits one, two, three, etc. places to the right and put in zeros if necessary. For example:

17.6 ÷ 10 = 1.76 — Move the digits one place to the right.
74 ÷ 100 = 0.74 — Move the digits two places to the right.
19.6 ÷ 1000 = 0.0196 — Move the digits three places to the right.

When dividing by multiples of 10, the same rules apply, except you divide the numbers and then move the digits to the right. For example:

6000 ÷ 30
= 6000 ÷ 3 ÷ 10
= 2000 ÷ 10
= 200

6.3 ÷ 30
= 6.3 ÷ 3 ÷ 10
= 2.1 ÷ 10
= 0.21

Long multiplication

When you multiply two or more numbers together you are finding the **product**.

Example

A tin of soup costs 64p. Work out the cost of 127 tins of soup.

Cost = 127 × 64

```
   127
    64 ×
  ─────
   508    ① 127 × 4
   1 2
  7620 +  ② 127 × 60
   1 4
  ─────
  8128    ③ 508 + 7620
     1
```

Cost = 8128p or £81.28

Alternatively, you can use a 'grid' method.

×	100	20	7	
60	6000	1200	420	→ 7620
4	400	80	28	→ 508 +

$$8128$$
$$1$$

= £81.28

Long division

The following example will help you to understand long division.

Example

A pencil costs 27p. Samuel has £3.66 to spend. What is the maximum number of pencils he can buy? How much change will he have left over?

```
     13
   ─────
27 )366
    27 ↓ −
   ─────
     96
     81 −
   ─────
     15
```

① 27 goes into 36 once, put down 1
② Place 27 below 36
③ Subtract 27 from 36 (get 9)
④ Bring down the 6
⑤ Divide 27 into 96, put down 3, since 3 × 27 = 81
⑥ Write down 81
⑦ 96 − 81 = remainder 15

Samuel can buy 13 pencils and he will have 15p left over.

The method of **chunking** can also be used when dividing. Always try to estimate the answer to your division.

```
27 )366
    270 −   27 × 10
   ─────
     96
     81 −   27 × 3
   ─────
     15
```

366 ÷ 27 = 10 + 3, remainder 15
= 13 pencils and 15p left over

Quick test

Answer the following questions, without using a calculator.

① a) 479 + 698   b) 379 − 147   c) 287 × 6   d) 5)1375

② 427 × 48

③ 49)1813

④ a) 16.4 × 10   b) 8.9 × 100   c) 37 × 1000   d) 97 ÷ 1000

⑤ The cost of a trip is £15.65. If Ms Kier collects £657.30, how many people are going on the trip?

9

Fractions

Fractions

A fraction is a part of a whole.
The top number is the numerator.
The bottom number is the denominator.

A fraction like $\frac{4}{5}$ is called a **proper fraction** because the **denominator is greater** than the numerator.

A fraction like $\frac{24}{17}$ is called an **improper fraction** because the **numerator is greater** than the denominator.

A number like $2\frac{1}{2}$ is called a **mixed number**.

When expressing one amount as a fraction of another amount, write the numbers as a fraction with the first amount the numerator and the second amount the denominator. For example, 7 as a fraction of 9 is written as $\frac{7}{9}$

Equivalent fractions

Equivalent fractions have the same value.
For example:

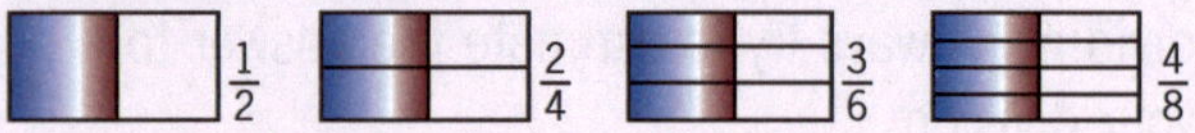

$\frac{1}{2}$ $\frac{2}{4}$ $\frac{3}{6}$ $\frac{4}{8}$

From the diagrams it can be seen that
$\frac{1}{2} = \frac{2}{4} = \frac{3}{6} = \frac{4}{8}$

They are equivalent fractions. Fractions can be changed to their equivalents by **multiplying** or **dividing** both the numerator and denominator by the same amount.

Examples
a) Change $\frac{5}{7}$ to its equivalent fraction with a denominator of 28.

Multiply the top and bottom by 4.
So $\frac{5}{7}$ is equivalent to $\frac{20}{28}$

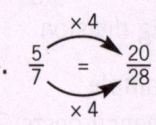

$\frac{5}{7} \xrightarrow{\times 4} = \frac{20}{28}$

b) Change $\frac{40}{60}$ to its equivalent fraction with a denominator of 3.

Divide the top and bottom by 20.
So $\frac{40}{60}$ is equivalent to $\frac{2}{3}$
$\frac{40}{60} = \frac{2}{3}$ in its simplest form.

$\frac{40}{60} \xrightarrow{\div 20} = \frac{2}{3}$

This is known as simplifying the fraction by '**cancelling**'.

Addition and subtraction of fractions

These examples show the basic principles of adding and subtracting fractions.

Examples
a) Work out $\frac{1}{8} + \frac{3}{4}$

❶ First make the denominators the same: $\frac{3}{4}$ is **equivalent** to $\frac{6}{8}$
❷ Replace $\frac{3}{4}$ with $\frac{6}{8}$ to make $\frac{1}{8} + \frac{6}{8}$
❸ Add the numerators: $1 + 6 = 7$
Do not add the denominators; the denominator stays the same.
So, $\frac{1}{8} + \frac{6}{8} = \frac{7}{8}$

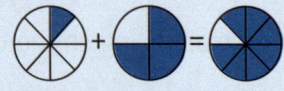

$\frac{3}{4} \xrightarrow{\times 2} = \frac{6}{8}$

b) Work out $\frac{9}{12} - \frac{1}{3}$

❶ First make the denominators the same: $\frac{1}{3}$ is equivalent to $\frac{4}{12}$
❷ Replace $\frac{1}{3}$ with $\frac{4}{12}$
❸ Subtract the numerators but **not** the denominators; the denominator stays the same.
So, $\frac{9}{12} - \frac{4}{12} = \frac{5}{12}$
Or, $\frac{9-4}{12} = \frac{5}{12}$

$\frac{1}{3} \xrightarrow{\times 4} = \frac{4}{12}$

Multiplication and division of fractions

When multiplying and dividing fractions, write out whole or mixed numbers as improper fractions before starting. This is an example of multiplication:

$\frac{2}{9} \times \frac{4}{7} = \frac{2 \times 4}{9 \times 7} = \frac{8}{63}$ ← Multiply numerators together. Multiply denominators together.

Change a division into a multiplication by turning the second fraction upside down and multiplying both fractions together; that is, **multiply by the reciprocal**.

Example
a) Work out $\frac{7}{9} \div \frac{12}{18}$

1. Take the **reciprocal** of the **second fraction**.
2. Multiply the fractions as normal.
   So, $\frac{7}{9} \times \frac{18}{12} = \frac{126}{108}$
3. Rewrite the answer as a mixed number.
   $\frac{126}{108} = 1\frac{18}{108} = 1\frac{1}{6}$

b) Work out $\frac{2}{3} \div \frac{1}{9} = \frac{2}{3} \times \frac{9}{1} = 6$

Using the fraction key on a calculator

$\boxed{a^{b/c}}$ is the fraction key on many calculators. For example, 12 out of 18 can be written as $\frac{12}{18}$. $\frac{12}{18}$ may be keyed as $\boxed{1}$ $\boxed{2}$ $\boxed{a^{b/c}}$ $\boxed{1}$ $\boxed{8}$

The display may be $\boxed{12 \lrcorner 18}$ or $\boxed{12 \ulcorner 18}$

The calculator will automatically cancel down fractions when the $\boxed{=}$ key is pressed.

For example, $\frac{12}{18}$ becomes $\boxed{2 \lrcorner 3}$ or $\boxed{2 \ulcorner 3}$. This means two-thirds.

A display of $\boxed{1 \lrcorner 4 \lrcorner 9}$ means $1\frac{4}{9}$. If you press $\boxed{shift}$ $\boxed{a^{b/c}}$, it converts to and from the improper fraction, $\boxed{13 \lrcorner 9}$, which means $\frac{13}{9}$.

💡 *Make sure you know how to use the fraction key on your calculator.*

Proportional changes with fractions

Increase and decrease
There are two methods for working out proportional changes with fractions. Use the one that is familiar to you. Remember that 'of' means multiply.

Example
Last year a gym had 290 members. This year there are $\frac{3}{5}$ more. How many members are there now?

Method 1
1. Work out $\frac{1}{5}$ of 290 = 58, so $\frac{3}{5} \times 290 = 174$
2. Add this to the original number, so
   290 + 174 = 464 people

Method 2
Increasing by $\frac{3}{5}$ is the same as multiplying by $1\frac{3}{5}$ $(1 + \frac{3}{5})$.
On the calculator, key in

$\boxed{1}$ $\boxed{a^{b/c}}$ $\boxed{3}$ $\boxed{a^{b/c}}$ $\boxed{5}$ $\boxed{\times}$ $\boxed{290}$ $\boxed{=}$

$1\frac{3}{5} \times 290 = 464$

💡 *Make sure you know how to use your calculator to add, subtract, multiply and divide fractions.*

Quick test

1. Without using a calculator, work out the following:
   a) $\frac{2}{9} + \frac{3}{27}$
   b) $\frac{3}{5} - \frac{1}{4}$
   c) $\frac{6}{9} \times \frac{72}{104}$
   d) $\frac{8}{9} \div \frac{2}{3}$
   e) $\frac{4}{7} - \frac{1}{3}$
   f) $\frac{2}{7} \div 1\frac{1}{2}$
   g) $\frac{7}{11} \div \frac{22}{14}$
   h) $\frac{2}{9} + \frac{4}{7}$

2. Calculate $\frac{2}{9}$ of £180.

3. $\frac{7}{12}$ more rain fell this year than last year. If 156mm fell last year, how much rain fell this year? 🖩

4. A mixed selection of biscuits has $\frac{1}{3}$ chocolate and $\frac{2}{5}$ jam cream biscuits. The rest of the biscuits are plain. What fraction of the biscuits are plain?

Decimals

Types of decimals

A **decimal point** is used to separate whole number columns from fractional columns.
For example:

Thousands	Hundreds	Tens	Units	.	Tenths	Hundredths	Thousandths
8	9	2	4	.	1	6	3

The 1 means $\frac{1}{10}$, the 6 means $\frac{6}{100}$, the 3 means $\frac{3}{1000}$

Terminating and recurring decimals
A decimal that stops is called a terminating decimal. All terminating decimals can be written as a fraction. For example:

$0.273 = \frac{273}{1000}$

$0.49 = \frac{49}{100}$

$0.7 = \frac{7}{10}$

If a fraction (in its simplest form) has a denominator with prime factors of **only** 2 and 5 then it will convert to a terminating decimal. Any other fraction will become a recurring decimal. A decimal that recurrs is shown by placing a dot over the numbers that repeat. For example:

$0.333... = 0.\dot{3}$
$0.232323... = 0.\dot{2}\dot{3}$
$0.1234123412 = 0.\dot{1}23\dot{4}$

Multiplying and dividing decimals by 10, 100, 1000

To multiply a decimal number by 10, 100, 1000, etc. move each digit one, two, three, etc. places to the left. For example:
8.7 × 10 = 87
15.62 × 10 = 156.2
15.27 × 100 = 1527
29.369 × 1000 = 29369

To divide decimals by 10, 100, 1000, etc. move each digit one, two, three, etc. places to the right. For example:
146.2 ÷ 100 = 1.462
27.6 ÷ 1000 = 0.0276

Ordering decimals

When ordering decimals, follow these steps:
❶ First write them with the same number of figures after the decimal point.
❷ Then compare whole numbers, digits in the tenths place, digits in the hundredths place, and so on.

Remember, hundredths are smaller than tenths: $\frac{10}{100} = \frac{1}{10}$ so $\frac{8}{100} < \frac{1}{10}$

Example
The results below show the distance, in metres, that some students achieved when throwing the discus:
16.21m, 16.02m, 16.4m, 16.04m, 12.71m, 18.3m
Arrange these distances in order of size, smallest first.

❶ Rewrite them with two figures after the decimal point: 16.21, 16.02, 16.40, 16.04, 12.71, 18.30
❷ Now reorder them:
12.71, 16.02, 16.04, 16.21, 16.40, 18.30
❸ Rewrite them in their original form: 12.71m, 16.02m, 16.04m, 16.21m, 16.4m, 18.3m

Multiplying and dividing by numbers between 0 and 1

When **multiplying** by numbers between 0 and 1, the result is always **smaller** than the starting value.

For example: 6 × 0.1 = 0.6
6 × 0.01 = 0.06
6 × 0.001 = 0.006

The results are all smaller than the starting values, in this case 6.

When **dividing** by numbers between 0 and 1, the result is always **bigger** than the starting value.

For example: 6 ÷ 0.1 = 60
6 ÷ 0.01 = 600
6 ÷ 0.001 = 6000

The results are all bigger than the starting values, in this case 6.

Calculations with decimals

When **adding** and **subtracting** decimals, the decimal points need to go under each other. For example:

27.46
7.291 +
─────
34.751

17.00
12.84 −
─────
4.16

Line up the digits carefully. Put the decimal points under each other. The decimal point in the answer will then be in line.

💡 *Remember to line up the digits carefully.*

When **multiplying** decimals, the answer must have the same number of decimal places as the total number of decimal places in the numbers that are being multiplied. For example, 84 × 1.4 = 117.6

Examples
a) Work out 24.6 × 7

246
7 ×
────
1722

Multiply 246 by 7 = 1722, ignoring the decimal point. 24.6 has 1 number after the decimal point. The answer must have 1 decimal place (1 d.p.).

So 24.6 × 7 = 172.2

b) Work out 4.52 × 0.2

452
2 ×
───
904

Work out 452 × 2, ignoring the decimal points. 4.52 has 2 d.p.; 0.2 has 1 d.p. So the answer must have 3 d.p.

904 → 0.904   *Move the digits 3 places to the right.*

So 4.52 × 0.2 = 0.904

When **dividing** decimals, divide as normal, placing the decimal points in line. For example:

    4.8
3)14.4   *Put the decimal points in line.*

A fraction can be changed into a decimal by dividing the numerator by the denominator. For example, this is how you would change $\frac{3}{5}$ into a decimal by short division:

    0.6
5)3.0

When dividing by a decimal, it is easier to multiply the numerator and denominator by a power of 10, so that it becomes equivalent to a division with a whole number. For example:

$$\frac{275}{0.25} = \frac{27\,500}{25}$$

Multiply the numerator and denominator by 100.

= 1100

Quick test

1 Without using a calculator, work out the following:
   a) 27.16 + 9.32   b) 29.04 − 11.361   c) 12.8 × 2.1   d) 49.2 ÷ 4
   e) 600 × 0.01   f) 520 × 0.1   g) 20 × 0.02   h) 37 × 0.0001
   i) 400 ÷ 0.1   j) 450 ÷ 0.01   k) 470 ÷ 0.001   l) 650 ÷ 0.02
   m) 6.93 × 100   n) 29.1 × 10   o) 707.4 ÷ 100   p) 28.4 ÷ 1000

2 Arrange these numbers in order of size, smallest first.
   a) 2.61, 4.02, 4.20, 4.021, 2.615, 2.607
   b) 8.27, 8.206, 8.271, 6.49, 6.05, 8.93

Rounding

Rounding to the nearest 10, 100, 1000

Large numbers are often rounded to the nearest ten, hundred or thousand to make them easier to work with.

Rounding to the nearest ten
Look at the **digit** in the units column. If it is **less than 5, round down**. If it is **5 or more, round up**.

Example
276 people went to the theatre to see the opening night of a play. Round this to the nearest ten.

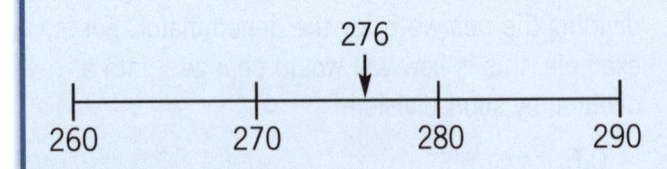

```
         276
          ↓
|----|----|----|----|
260  270  280  290
```

There is a 6 in the units column.
6 > 5, so round up to 280.
276 is 280 to the nearest ten.

Rounding to the nearest hundred
Look at the digit in the tens column. If it is less than 5, round down. If it is 5 or more, round up.

Example
One day in July, 3250 people went to a theme park. Round this to the nearest hundred.

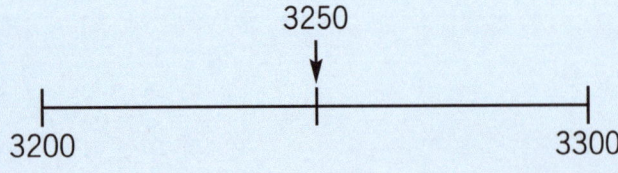

```
      3250
       ↓
|------|------|
3200        3300
```

Since there is a 5 in the tens column, you round up to 3300.
3250 is 3300 to the nearest hundred.

Rounding to the nearest thousand
Look at the digits in the hundreds column. The same rules apply as for rounding to the nearest ten or hundred.

Example
17 436 people attended a football match. Round this to the nearest thousand.

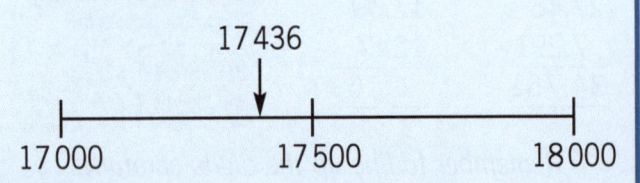

```
        17 436
          ↓
|----------|----------|
17 000   17 500    18 000
```

There is a 4 in the hundreds column, so round down to 17 000.
17 436 is 17 000 to the nearest thousand.

Similar methods can be used to round any number to any power of 10.

Rounded numbers are often used in newspaper reports.

Decimal places (d.p.)

When rounding numbers to a specified number of **decimal places**, follow these steps:

1. Look at the last number that is wanted (e.g. if rounding 12.367 to 2 decimal places, look at the 6 which is the second decimal place).
2. Look at the number to the right of it (the number that is not needed – in this case, the 7).
3. If it is **5 or more**, then **round up** the last digit (7 is greater than 5, so round up the 6 to a 7).
4. If it is **less than 5**, then the last digit remains the **same**.

In athletics, times are usually rounded to 2 decimal places. The men's 100m world record holder in 2009 was Usain Bolt, of Jamaica, with a time of 9.58 seconds (2 d.p.).

Examples

a) Round 12.49 to 1 decimal place.

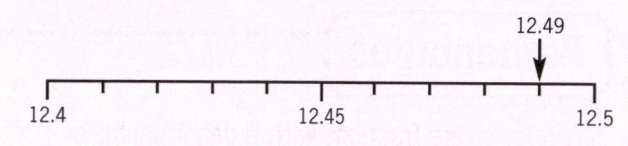

12.49 rounds up to 12.5

b) Round 8.735 to 2 decimal places.

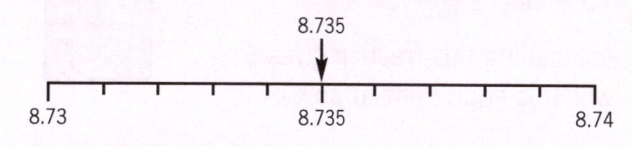

8.735 rounds up to 8.74

c) Round 9.624 to 2 decimal places.

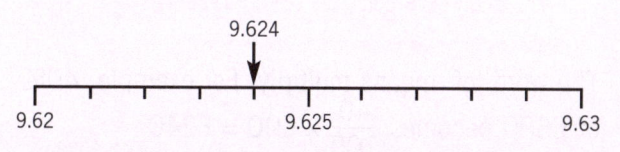

9.624 rounds down to 9.62

Significant figures (s.f. or sig. fig.)

The first **significant figure** is the first digit that is not zero. The 2nd, 3rd, 4th... significant figures follow on after the first digit. They may or may not be zeros. For example:

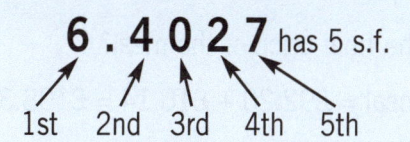

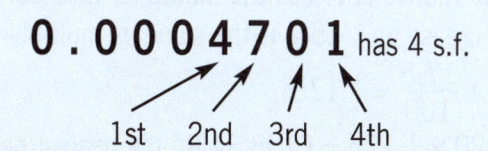

6.4027 has 5 s.f.
1st 2nd 3rd 4th 5th

0.0004701 has 4 s.f.
1st 2nd 3rd 4th

To round a number to a given number of significant places, apply the same rule as with decimal places: if the next digit is 5 or more, round up.

For example:

Number	to 3 s.f.	to 2 s.f.	to 1 s.f.
4.207	4.21	4.2	4
4379	4380	4400	4000
0.006 209	0.006 21	0.0062	0.006

After rounding the last digit, you must fill in the end zeros.
For example, 4380 = 4400 to 2 s.f. (not 44).

In 2009, the tallest structure in the world was the Burj Khalifa at a height of 828m; that is 800m correct to 1 s.f.

Take care when rounding that you do not change the place values.

Quick test

1. Round the following numbers to 2 decimal places.
   a) 6.429   b) 18.607   c) 14.271   d) 29.638
2. The length of the River Nile is 6650km. What is this length correct to 2 significant figures?

Percentages 1

Percentages

Percentages are fractions with a denominator of 100.

% is the percentage sign.

75% means $\frac{75}{100}$

(cancelling this fraction gives $\frac{3}{4}$ which is equivalent to 75%).

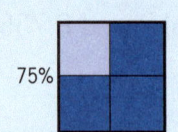

75%

Example
A mixed bag of nuts contains 26% cashew nuts, 42% peanuts and the rest are pistachio nuts. What percentage are pistachio nuts?

26 + 42 = 68% cashew nuts and peanuts.
Percentage of pistachio nuts = 100% − 68% = 32%

Percentage of a quantity

The word '**of**' means **multiply**. For example, 40% of £600 becomes $\frac{40}{100} \times 600 = £240$

On the calculator, key in

| 40 | ÷ | 100 | × | 600 | = |

To work this out without a calculator, follow these steps:

① Work out 10% first by dividing by 10, i.e. 600 ÷ 10 = £60
② Multiply by 4 to get 40%, i.e. 4 × 60 = £240

Examples

a) Find 15% of £750, without using a calculator.

$10\% = \frac{1}{10}$ so 10% of £750 = $\frac{750}{10}$ = £75

5% of 750 is half of 75 = £37.50
So 15% = £75 + £37.50 = £112.50

b) Find 17.5% of £640, without using a calculator.

10%   = £640 ÷ 10 = £64
5%   = £32
2.5% = £16
So 17.5% = £64 + £32 + £16 = £112

c) A jumper is reduced in a sale by 15%. The original price of the jumper was £42. Work out the sale price, without using a calculator.

15% of £42
10% = £42 ÷ 10 = £4.20
5% = £4.20 ÷ 2 = £2.10
So 15% = £4.20 + £2.10 = £6.30
Sale price of jumper is £42 − £6.30 = £35.70

d) A meal for four costs £92.20. VAT (Value Added Tax), a tax that is added on to the cost of most items, is charged at 17.5%.

i) How much VAT is there to pay on the meal?

17.5% of £92.20 = $\frac{17.5}{100} \times 92.20$
= £16.14 (to the nearest penny)
VAT = £16.14

This is just like a 'percentage of a quantity' question.

ii) What is the final price of the meal?

Price of meal = £92.20 + £16.14 = £108.34

An alternative is to use the multiplier method:
An increase of 17.5% is the same as multiplying by $1 + \frac{17.5}{100} = 1.175$

£92.20 × 1.175 = £108.34 (to the nearest penny)

e) A sofa is reduced in a sale by 30%. The original price of the sofa was £825. Work out the sale price of the sofa.

Using the multiplier method:
A decrease of 30% is the same as multiplying by $1 - \frac{30}{100} = 0.7$

£825 × 0.7 = £577.50

Percentage increase and decrease

To write one amount as a percentage of another, first find the fraction then multiply by 100.

$$\% \text{ change} = \frac{\text{change}}{\text{original}} \times 100\%$$

Example
A Maths test score is 13 out of 20.

$$\frac{13}{20} \times 100$$
$$= 65\%$$

Example
A coat costs £125. In a sale the price is reduced to £85.
What is the percentage reduction?

Reduction = £125 − £85 = £40
$$= \frac{40}{125} \times 100\%$$
$$= 32\%$$

Example
Matthew bought a flat for £45 000.
Three years later, he sold it for £62 000.
What was his percentage profit?

Profit = £62 000 − £45 000
= £17 000

% Profit = $\frac{17\,000}{45\,000} \times 100\%$
= 37.78% (2 d.p.)

Example
Ahmed sold his mp3 player for £30. He had bought it for £59. What is his percentage loss?

Loss = £59 − £30
= £29
= $\frac{29}{59} \times 100\%$
= 49% (2 s.f.)

Quick test

1. Work out 30% of £700

2. Sarah got 94 out of 126 in a maths test. What percentage did she get? Give your answer to 1 decimal place.

3. Reece weighed 6lb when he was born. If his weight has increased by 65%, how much does he now weigh? Give your answer to 1 decimal place.

4.
Super's football boots	Joe's football boots
$\frac{1}{3}$ off	28% off

If a pair of football boots usually costs £49.99, which shop sells them cheaper in the sale and what is the sale price?

Percentages 2

Repeated percentage change

Example
A car was bought for £8000. Each year it depreciates in value by 20%.
What is the car worth 3 years later?

Method 1
- First find 80% of the value of the car.

  Year 1 $\dfrac{80}{100} \times £8000 = £6400$

- Then work out the value year by year.

  Year 2 $\dfrac{80}{100} \times £6400 = £5120$

  (£6400 depreciated in value by 20%)

  Year 3 $\dfrac{80}{100} \times £5120 = £4096$ after 3 years

  (£5120 depreciated by 20%)

Work these questions out year by year.

Beware! Do not do 3 × 20 = 60% reduction over 3 years!

Method 2
A quick way to work this out uses the **multiplier** method.

- Finding 80% of the value of the car is the same as multiplying by 0.8. The scale factor is 0.8.

  Year 1   0.8 × £8000 = £6400
  Year 2   0.8 × £6400 = £5120
  Year 3   0.8 × £5120 = £4096

This is the same as working out
$(0.8)^3 \times 8000 = £4096$

This is a much quicker way if you understand it; and also it's useful to know for everyday life.

Simple interest

Simple interest is the interest that is sometimes paid on money in banks and building societies. The interest is paid each year (**per annum** or **p.a.**) and is the same amount each year.

Example
Jonathan has £2500 in his savings account. **Simple interest** is paid at 4.4% p.a. How much does he have in his account at the end of the year?
(This is a 'percentage of' question.)

100 + 4.4 = 104.4% (increasing by 4.4% is the same as multiplying by 100 + 4.4 = 104.4%)

Total savings = $\dfrac{104.4}{100} \times £2500 = £2610$

Note: If the money was in the account for 4 years, the interest at the end of the 4 years would be 4 × £110 = £440.

Interest paid = £2610 − £2500 = £110

Compound interest

Compound interest is where banks pay interest on the interest earned as well as on the original sum.

Example
If Jonathan has £2500 in his savings account and **compound interest** is paid at 4.4% p.a. how much will he have in his account after 2 years?

Method 1
Year 1: $\dfrac{104.4}{100} \times £2500 = £2610$
Year 2: $1.044 \times £2610 = £2724.84$
Total after 2 years = £2724.84

Method 2
Using the multiplier method:
$\dfrac{104.4}{100} = 1.044$ is the scale factor

£2500 × 1.044 × 1.044
= 2500 × $(1.044)^2$

Total after 2 years = £2724.84

National Insurance and Tax

National Insurance
National Insurance (NI) is usually deducted as a percentage from a wage or salary.

Example
Sue earns £1402.65 a month. National Insurance at 11% is deducted. How much NI does she pay?

11% of £1402.65 = 0.11 × £1402.65
= £154.29

Income tax
A percentage of a wage or salary is deducted as **income tax**. **Personal allowances** must first be deducted in order to obtain the **taxable income**.

Example
Harold earns £190 per week. The first £62 is not taxable but the remainder is taxed at 20%. How much income tax does he pay each week?

Taxable income = £190 − £62 = £128
20% tax = 0.2 × £128 = £25.60
Tax per week = £25.60

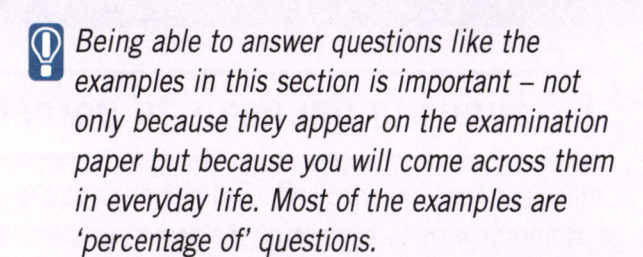

Being able to answer questions like the examples in this section is important – not only because they appear on the examination paper but because you will come across them in everyday life. Most of the examples are 'percentage of' questions.

There are really only 2 types of percentage questions:
- *'Percentage of'. Here you are given the percentage so you divide by 100.*
- *Writing an answer as a percentage. Here you need to work out a fraction or decimal and multiply by 100.*

Quick test

1. Charlotte has £4250 in the bank. If the interest rate is 3.8% p.a. how much interest on the savings will she get at the end of the year?

2. A car costs £6000 cash, or can be bought by hire purchase with a 30% deposit followed by 12 monthly instalments of £365. Find…
   a) the deposit
   b) the total amount paid for the car on hire purchase.

3. A flat was bought for £162 000. A year later, the price increased by 20% and then by a further 5% the following year. How much was the flat worth at the end of the second year?

4. Fiona has £3200 in savings. If compound interest is paid at 3% p.a. how much will she have in her account after 2 years?

Fractions, decimals & percentages

Fractions to decimals to percentages

Equivalent fractions, decimals and percentages are all different ways of expressing the same number.

The table opposite shows...
- some common fractions and their **equivalents**, which you need to learn
- how to convert fractions to decimals to percentages.

Fraction	Decimal	Percentage
$\frac{1}{2}$	0.5	50%
$\frac{1}{3}$	$0.\dot{3}$	$33.\dot{3}\%$
$\frac{2}{3}$	$0.\dot{6}$	$66.\dot{6}\%$
$\frac{1}{4}$	0.25	25%
$\frac{3}{4}$	$\xrightarrow{3 \div 4}$ 0.75 $\xrightarrow{\times 100\%}$	75%
$\frac{1}{5}$	0.2	20%
$\frac{1}{8}$	0.125	12.5%
$\frac{3}{8}$	0.375	37.5%
$\frac{1}{10}$	0.1	10%
$\frac{1}{100}$	0.01	1%

Ordering fractions

When ordering fractions, it is useful to write them with a common denominator so that you can compare the numerators.

Example
Place these fractions in order of size, smallest first. $\frac{1}{4}, \frac{3}{10}, \frac{5}{8}, \frac{7}{20}, \frac{27}{40}$

❶ Since all the fractions can be written with a denominator of 40, we can compare the numerators.
$\frac{10}{40}, \frac{12}{40}, \frac{25}{40}, \frac{14}{40}, \frac{27}{40}$

❷ In size order:
$\frac{10}{40}, \frac{12}{40}, \frac{14}{40}, \frac{25}{40}, \frac{27}{40} = \frac{1}{4}, \frac{3}{10}, \frac{7}{20}, \frac{5}{8}, \frac{27}{40}$

Ordering different types of numbers

When putting a mixture of fractions, decimals and percentages in order of size, it is best to change them all to decimals first.

Example
Place the following in order of size, smallest first.
$\frac{3}{5}$, 0.65, 0.273, 27%, 62%, $\frac{4}{9}$

0.6, 0.65, 0.273, 0.27, 0.62, $0.\dot{4}$ *Put into decimals first.*

0.27, 0.273, $0.\dot{4}$, 0.6, 0.62, 0.65 *Place in order of size, smallest first.*

27%, 0.273, $\frac{4}{9}$, $\frac{3}{5}$, 62%, 0.65 *Rewrite.*

❗ Get a friend to test you on the equivalences between fractions, decimals and percentages because you need to know them.

Quick test

❶ Change the following fractions into **a)** decimals **b)** percentages.
i) $\frac{2}{7}$ ii) $\frac{3}{5}$ iii) $\frac{8}{9}$

❷ Place the following in order of size, smallest first.
$\frac{2}{5}$, 0.42, 0.041, $\frac{1}{3}$, 5%, 26%

Using a calculator

Order of operations

BIDMAS is a made-up word that helps you to remember the order in which calculations take place.

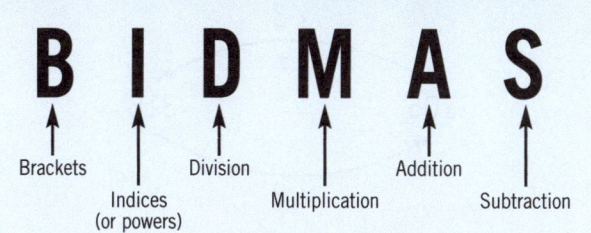

B I D M A S

- Brackets
- Indices (or powers)
- Division
- Multiplication
- Addition
- Subtraction

BIDMAS simply means that anything in brackets is done first, followed by indices, division and multiplication, then addition and subtraction.

For example:
$(5 + 2) \times 3 = 21$ — Do the calculation in brackets first: $5 + 2 = 7$, then $7 \times 3 = 21$

$5 + 2 \times 3 = 11$ — Here there are no brackets, so the multiplication (2×3) is carried out first, then $5 + 6 = 11$

Important calculator keys

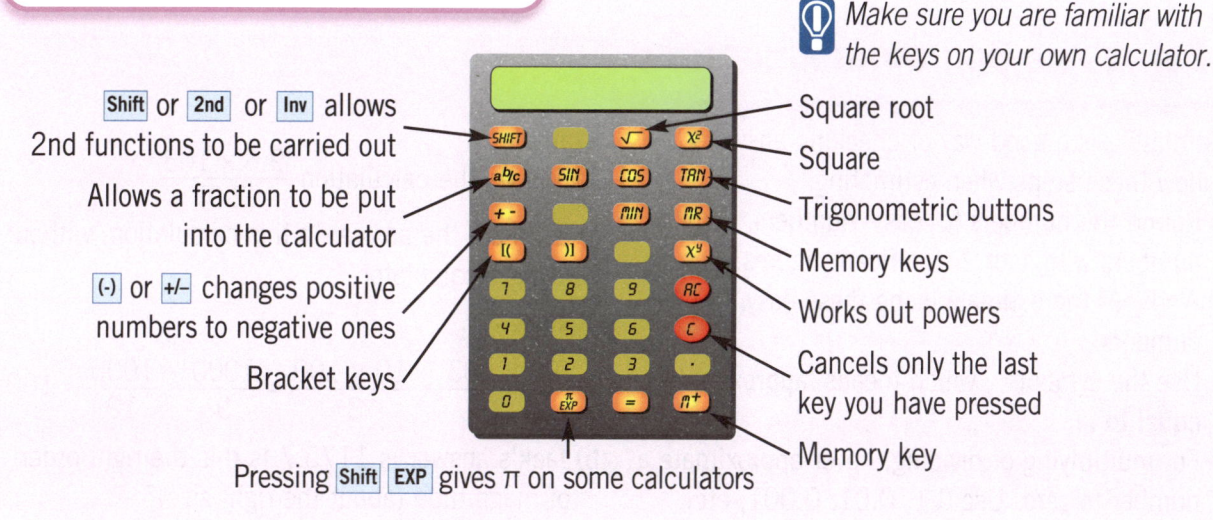

Make sure you are familiar with the keys on your own calculator.

- Shift or 2nd or Inv allows 2nd functions to be carried out
- Allows a fraction to be put into the calculator
- (-) or +/− changes positive numbers to negative ones
- Bracket keys
- Square root
- Square
- Trigonometric buttons
- Memory keys
- Works out powers
- Cancels only the last key you have pressed
- Memory key

Pressing Shift EXP gives π on some calculators

Calculating powers

y^x, x^y, $\wedge$ or $x^\square$ is used for calculating powers, such as 2^7.

Use the power key on your calculator to work out 2^7:
1. Write down the calculator keys used.
2. Check that you obtain the answer 128.

Write down the keys that are needed for the following calculations. Work the calculations out in stages. Check that you get the right answers.

a) $\dfrac{2.9 \times 3.6}{(4.2 + 3.7)} = 1.322$ (3 d.p.) ← $\dfrac{10.44}{7.9} = 1.322$ (2 d.p.)

b) $9^2 \times 4^5 = 82944$

c) $\dfrac{3 \times (5.2)^2}{(2.3)^3 \times 5} = 1.33$ (2 d.p.) ← $\dfrac{81.12}{60.835} = 1.33$ (2 d.p.)

Make sure you know how to use the power key. It can save lots of time.

Quick test

1. Work out the following on your calculator. Round your answers to 2 d.p.

   a) $\dfrac{27.1 \times 6.4}{9.3 + 2.7}$

   b) $\dfrac{(9.3)^4}{2.7 \times 3.6}$

   c) $\sqrt{\dfrac{25^2}{4\pi}}$

   d) $\dfrac{5}{9}(25 - 10)$

Approximations & checking calculations

Checking calculations

When checking calculations, the process can be reversed. For example:

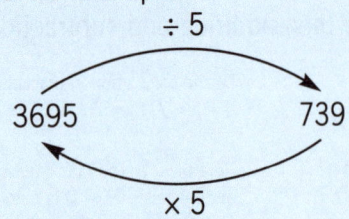

$3695 \xrightarrow{\div 5} 739$
$739 \xrightarrow{\times 5} 3695$

$3695 \div 5 = 739$
Check: $739 \times 5 = 3695$

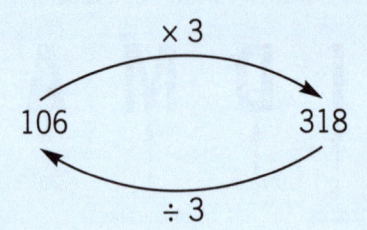

$106 \xrightarrow{\times 3} 318$
$318 \xrightarrow{\div 3} 106$

$106 \times 3 = 318$
Check: $318 \div 3 = 106$

Estimates and approximations

Estimating is a good way of checking answers. Follow these steps when estimating:

1. Round the numbers to 'easy' numbers, usually numbers with 1 or 2 significant figures.
2. Work out the estimate using these easy numbers.
3. Use the symbol ≈, which means '**approximately equal to**'.
4. For multiplying or dividing, never approximate a number to zero. Use 0.1, 0.01, 0.001, etc.

For example:

a) $8.93 \times 25.09 \approx 10 \times 25 = 250$

b) $(6.29)^2 \approx 6^2 = 36$

c) $\dfrac{296 \times 52.1}{9.72 \times 1.14} \approx \dfrac{300 \times 50}{10 \times 1} = \dfrac{15000}{10} = 1500$

d) $0.096 \times 79.2 \approx 0.1 \times 80 = 8$

e) $\dfrac{602 \times 39}{0.213} \approx \dfrac{600 \times 40}{0.2}$

$= \dfrac{24000}{0.2}$

$= \dfrac{240000}{2}$ ← Multiply the numerator and denominator by 10.

$= 120000$

Example
Jack does the calculation $\dfrac{9.6 \times 103}{(2.9)^2}$

a) Estimate the answer to this calculation, without using a calculator.

Estimate:
$\dfrac{9.6 \times 103}{(2.9)^2} \approx \dfrac{10 \times 100}{3^2} = \dfrac{1000}{9} \approx \dfrac{1000}{10} = 100$

b) Jack's answer is 1175.7 Is this the right order of magnitude (about the right size)?

Jack's answer is not the right order of magnitude. It is 10 times too big.

When adding and subtracting, very small numbers may be approximated to zero.

For example:

$109.6 + 0.0002 \approx 110 + 0 = 110$

$63.87 - 0.01 \approx 64 - 0 = 64$

💡 *Questions that involve approximating are common on the non-calculator paper. For most of these questions, you are expected to round to 1 significant figure. Even if you find the calculation difficult, show your approximations to pick up method marks.*

Calculations

When solving problems, the answers should be rounded sensibly.

For example:
95.26 × 6.39 = 608.7114 = 608.71 (2 d.p.)
(Round to 2 decimal places because the values in the question are to 2 decimal places.)

Examples

a) Jackie has £9.37. She divides it as equally as she can between 5 people. How much does each person receive?

£9.37 ÷ 5 = £1.874
= £1.87
(Round to 2 d.p. as it is money.)

> *You will lose marks if you do not write money to 2 d.p. For example, if the answer to a money calculation is £9.7, write it, to 2 d.p., as £9.70*

b) Paint is sold in 8-litre tins. Sandra needs 27 litres of paint. How many tins must she buy?

27 ÷ 8 = 3.375 or 3 remainder 3

Sandra needs 4 tins of paint.

Sandra would not have enough paint with 3 tins – she would be 3 litres short. Hence the number of tins of paint must be rounded up.

> *When rounding remainders, consider the context of the question.*

c) The price of a basic calculator is £2.30. Mr Baur wants to buy some basic calculators. He has £70 to spend. Work out the greatest number of basic calculators he can buy.

70 ÷ 2.30 = 30.434… calculators

The greatest number of basic calculators that Mr Baur can buy is 30.

d) Mrs Chan is tiling her bathroom. She needs 43 tiles. Tiles are sold in packs of 8 and cost £15.66 per pack. How much will it cost Mrs Chan to tile her bathroom?

Number of packs of tiles: 43 ÷ 8 = 5.375
She needs 6 packs of tiles.
Total cost = 6 × £15.66 = £93.96

Quick test

1. Estimate the answer to $\dfrac{(29.4)^2 + 106}{2.2 \times 5.1}$

2. Sukhvinder decided to decorate her living room. The total area of the walls was 48m^2. If one roll of wallpaper covers 5m^2 of wall, how many rolls of wallpaper did Sukhvinder need?

3. Thomas earned £171.35 for working a 23-hour week. How much was he paid per hour? Check your calculation by estimating.

Ratio

Ratios

A ratio is used to compare two or more related quantities:
- '**Compared to**' is replaced with **two dots** (:)
  For example, '16 boys compared to 20 girls' can be written as 16 : 20.
- To simplify ratios, divide both parts of the ratio by their highest common factor.
  For example, 16 : 20 = 4 : 5 *Divide both sides by 4.*

When ratios are simplified as much as they can be, they are in their **simplest form**.

They can also be written in the form $1 : n$.

Examples
a) Write the ratio 12 : 24 in the ratio $1 : n$

12 : 24 = 1 : 2 *Divide both sides by 12.*

b) Simplify the ratio 21 : 28

21 : 28 = 3 : 4 *Divide both sides by 7.*

Look at the flowers shown below. The ratio of red flowers to purple flowers can be written:

÷2 (10 : 4) ÷2
= 5 : 2

In other words, for every 5 red flowers there are 2 purple flowers.

Out of every 7 flowers, 5 are red and 2 are purple, so the ratio of 5 : 2 is the same as $\frac{5}{7} : \frac{2}{7}$

Sharing a quantity in a given ratio

When sharing a quantity in a given ratio, follow these steps:
1. Add up the total parts.
2. Work out what one part is worth.
3. Work out what the other parts are worth.

Example
£20 000 is shared between Ewan and Leroy in the ratio 1 : 4. How much does each receive?

1 + 4 = 5 parts
5 parts = £20 000, so 1 part = $\frac{£20\,000}{5}$ = £4000

So, Ewan gets 1 × £4000 = £4000 and Leroy gets 4 × £4000 = £16 000

Best buys

Compare **unit amounts** to decide which option is the better value for money.

Example
The same brand of coffee is sold in two different-sized jars. Which jar represents the better value for money?

Find the cost per gram for both jars.

100g costs 186p so 186 ÷ 100 = 1.86p per gram.
250g costs 397p so 397 ÷ 250 = 1.588p per gram.

Since the larger jar costs less per gram, it offers the better value for money.

For a non-calculator method, work out the cost of 50g for each jar:

100g = £1.86 so 50g = 186 ÷ 2 = 93p
250g = £3.97 so 50g = 397 ÷ 5 = 79.4p

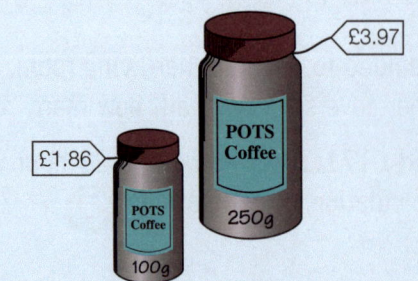

Hence, the larger jar is better value because it costs less per 50g.

Increasing and decreasing in a given ratio

The unitary method is useful when solving ratio problems:
1. Divide to get one part.
2. Multiply for each new part.

Examples

a) A photograph of length 9cm is to have its sides enlarged in the ratio 5 : 3. What is the length of the enlarged photograph?

1. Divide 9cm by 3 to get 1 part.
   $9 \div 3 = 3$
2. Multiply this by 5.
   $5 \times 3 = 15$cm

So the length 9cm becomes 15cm on the enlarged photograph.

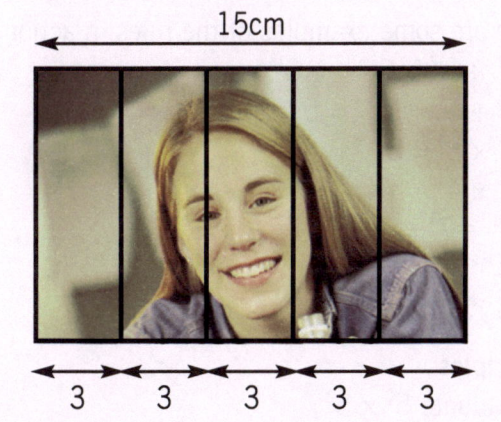

b) It took 8 people 6 days to build a wall. At the same rate, how long would it take 3 people?

Time for 8 people = 6 days
Time for 1 person = $8 \times 6 = 48$ days
(It takes 1 person longer to build the wall.)
3 people will take $\frac{1}{3}$ of the time taken by 1 person.

So time for 3 people = $\frac{48}{3}$ = 16 days

c) A recipe for 4 people needs 1600g of flour. How much flour is needed to make the recipe for 6 people?

1. Divide 1600g by 4, so 400g for 1 person.
2. Multiply by 6, so $6 \times 400g = 2400g$ of flour is needed for 6 people.

When answering problems of the type shown here, always try to work out what a unit (or one) is worth. You will then be able to work out what any other value is from that.

Quick test

1. Write the following ratios in their simplest form.
   a) 12 : 15   **b)** 6 : 12   **c)** 25 : 10

2. Three sisters share 60 sweets between them in the ratio 2 : 3 : 7. How many sweets does each sister receive?

3. If 15 oranges cost £1.80, how much will 23 of the same oranges cost?

4. The ratio of the lengths on a map and its enlargement are 7 : 12. If a road length was 21cm on the original map, what is the length of the road on the enlarged map?

25

Indices

Indices

An **index** (plural: **indices**) is sometimes known as a **power**.
For example:
- 6^4 is read as '6 to the power of 4'.
  It means $6 \times 6 \times 6 \times 6$.
- 2^7 is read as '2 to the power of 7'.
  It means $2 \times 2 \times 2 \times 2 \times 2 \times 2 \times 2$.

The **base** has to be the same when the rules of indices are applied.

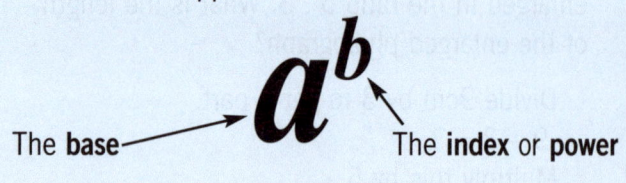

$$a^b$$

The **base** ⟶   ⟵ The **index** or **power**

Rules of indices

You need to learn the following rules.

Rule 1
When **multiplying**, **add** the powers.
$$4^7 \times 4^3 = 4^{7+3} = 4^{10}$$
Since:
$(4 \times 4 \times 4 \times 4 \times 4 \times 4 \times 4) \times (4 \times 4 \times 4) =$
$4 \times 4 \times 4 \times 4 \times 4 \times 4 \times 4 \times 4 \times 4 \times 4$

Rule 2
When **dividing**, **subtract** the powers.
$$6^9 \div 6^4 = 6^{9-4} = 6^5$$
Since:
$$\frac{\cancel{6} \times \cancel{6} \times \cancel{6} \times \cancel{6} \times 6 \times 6 \times 6 \times 6 \times 6}{\cancel{6} \times \cancel{6} \times \cancel{6} \times \cancel{6}}$$
$= 6 \times 6 \times 6 \times 6 \times 6$

Rule 3
Any number raised to the **power 0** is always **1**, provided the number is not 0.
$5^0 = 1$ $6^0 = 1$
$2.7189^0 = 1$ $0^0 =$ undefined

Rule 4
Anything to the **power 1** is just **itself**.
$15^1 = 15$ $1923^1 = 1923$

Rule 5
When raising one power to another, multiply the powers.
$$(3^2)^4 = 3^{2 \times 4} = 3^8$$
$$(6^3)^5 = 6^{3 \times 5} = 6^{15}$$

The above rules also apply when the powers are negative.

Here are some examples of the rules in action:
a) $6^2 \times 6^{12} = 6^{2+12} = 6^{14}$
b) $5^6 \div 5^4 = 5^{6-4} = 5^2$
c) $8^4 \times 8^3 = 8^{4+3} = 8^7$
d) $4^3 \times 4^{10} = 4^{3+10} = 4^{13}$
e) $7^{10} \div 7^4 = 7^{10-4} = 7^6$
f) $(4^3)^2 = 4^{3 \times 2} = 4^6$
g) $(5^7)^3 = 5^{7 \times 3} = 5^{21}$

Examples
a) Evaluate $5^1 \times 5^2$
$$5^1 \times 5^2 = 5^3 = 125$$

b) Evaluate $8^6 \div 8^4$
$$8^6 \div 8^4 = 8^2 = 64$$

❗ *Evaluate means 'work out'.*

Indices and algebra

The rules that apply with numbers also apply to algebra.

Laws of indices

$a^n \times a^m = a^{n+m}$
$a^n \div a^m = a^{n-m}$
$a^1 = a$
$a^0 = 1$
$(a^n)^m = a^{n \times m}$

Examples

a) Simplify $x^4 \times x^7$

$x^4 \times x^7 = x^{11}$

b) Simplify $a^{10} \div a^4$

$a^{10} \div a^4 = a^6$

c) Simplify $\dfrac{x^9}{x^4}$

$\dfrac{x^9}{x^4} = x^5$

d) Simplify $(a^4)^2$

$(a^4)^2 = a^{4 \times 2} = a^8$

e) Simplify $4x^2 \times 3x^5$

$4x^2 \times 3x^5 = 12x^7$

Note that the numbers are multiplied... ...but the powers of the same term are added.

f) Simplify $a^4 \times 3a^5$

$a^4 \times 3a^5 = 3a^9$

g) Simplify $12x^5 \div 2x^2$

$12x^5 \div 2x^2 = 6x^3$

Note that the numbers are divided... ...but the powers of the same term are subtracted.

h) Simplify $15p^9 \div 3p^4$

$15p^9 \div 3p^4 = 5p^5$

i) Simplify $\dfrac{3x^7 \times 4x^9}{6x^4}$

$\dfrac{3x^7 \times 4x^9}{6x^4} = \dfrac{12x^{16}}{6x^4} = 2x^{12}$

Work this out in two stages.

j) Simplify $x^6 \times 4x^3$

$x^6 \times 4x^3 = 4x^9$

k) Simplify $\dfrac{12a^2b^3}{6a^3b^2}$

$\dfrac{12a^2b^3}{6a^3b^2} = \dfrac{2b}{a}$

Divide the numbers: $12 \div 6 = 2$
Use the laws of indices to simplify each term.

l) Simplify $\dfrac{4a^4b^3}{2ab}$

$\dfrac{4a^4b^3}{2ab} = 2a^3b^2$

m) Simplify $(2x^3)^2$

$(2x^3)^2 = 2x^3 \times 2x^3$
$= 4x^6$

Either write out fully or... ...use the laws of indices.

or $(2x^3)^2 = 2^2 x^{3 \times 2}$
$= 4x^6$

n) Simplify $(3x^2)^3$

$(3x^2)^3 = (3x^2) \times (3x^2) \times (3x^2)$
$= 27x^6$

or $(3x^2)^3 = 3^3 x^{2 \times 3}$
$= 27x^6$

> Learn the laws of indices. Work out the more difficult questions in two stages.

Quick test

1 Simplify the following:
a) $12^4 \times 12^8$ b) $9^2 \times 9^4$ c) 4^1 d) $18^6 \div 18^2$ e) $4^7 \times 4^2$ f) 1^{20}

2 Simplify the following:
a) $x^4 \times x^9$ b) $2x^9 \times 3x^7$ c) $12x^4 \div 3x^2$ d) $\dfrac{25x^9}{5x^2}$ e) $\dfrac{5x^6 \times 4x^9}{10x^3}$ f) $(4x^5)^2$

Practice questions

Use these questions to test your progress. Check your answers on page 94. You may wish to answer these questions on a separate piece of paper so that you can show full working out, which you will be expected to do in the exam.

1 From this list of numbers (2, 9, 21, 40, 41, 64, 100):

    a) Write down the numbers that are odd

    b) Write down the square numbers

    c) Write down the prime numbers

    d) Write down any numbers that are factors of 80

    e) Write down any numbers that are multiples of 4

2 The temperature outside is -6°C, inside it is 15 degrees higher. What is the temperature inside?

3 At a school charity concert, £1674 is raised with the sale of 62 tickets. Work out the price of each ticket.

4 Write these numbers in figures.
    a) Fifty-six   **b)** Five hundred and eight   **c)** Seven thousand and two

5 Put these numbers in order of size, smallest to largest.

    639   728   405   736   829   27

6 Hussain scored 58 marks out of 75 in a test. What percentage did he get?

7 A school raises £525 at the summer fair. 60% of the money raised is used to repair the tennis courts. How much is used to repair the tennis courts?

8 Arrange these decimals in order of size, smallest to largest.
    6.39   5.42   5.04   6.27   6.385   5.032

9 Write the ratio 12 : 15 in its simplest form.

10) Emily and Matthew want to have a night out. They can go bowling, to the cinema or a concert. If they go to the concert they will walk there and back. If they go bowling or to the cinema they will need to travel by taxi each way.

	Distance from home	Ticket price per person
Bowling	18 miles	£4.45
Cinema	13 miles	£7.50
Concert	Walking distance	£49.50

There are two taxi companies they can choose:

	Charge per journey
John's taxi	£3.20 per mile
Robert's cars	£2.50 per mile plus £15

Use this information to work out the cheapest night out.

11) Toothpaste is sold in three different-sized tubes.
   50ml = £1.24   75ml = £1.96   100ml = £2.42
Which of the tubes of toothpaste is the best value for money? You must show full working out in order to justify your answer.

12) A piece of writing paper is 0.01cm thick. A notepad has 150 sheets of paper. How thick is the notepad?

13) A car was bought in 2004 for £9000. Each year it depreciates in value by 15%. What was the car worth two years later?

14) Simplify the following:
   a) $2^3 \times 2^2$ **b)** $a^4 \times a^3$ **c)** $b^6 \div b^2$ **d)** $\dfrac{2a^5 \times 8a^4}{4a^2}$

15) The table shows the cost of a single room at a hotel.

	Cost per person per night excluding VAT	
Day	Friday to Sunday	Monday to Thursday
Low season	£65.00	£95.00
High season	£90.00	£125.00

Frances stays at the hotel for three nights in high season. She arrives on Tuesday. How much does Frances pay for her room including VAT at 17.5%?

How well did you do?

| 0–4 | Try again | 5–7 | Getting there | 8–10 | Good work | 11–15 | Excellent! |

Algebra

Algebraic conventions

Make sure you understand the following conventions of algebra:
- A **term** is a collection of numbers, letters and brackets, all multiplied together.
- Terms are separated by + and − signs. Each term has a + or − attached to the front of it.
- An **expression** is a group of terms, e.g.:

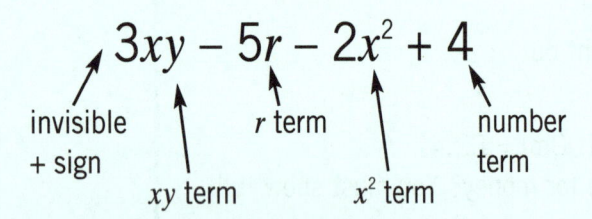

$$3xy - 5r - 2x^2 + 4$$

invisible + sign
xy term
r term
x^2 term
number term

- An algebraic expression must contain at least one letter.
- $5 \times a$ is written without the multiplication sign as $5a$.

$a + a + a = 3a$
$a \times a \times 2 = 2a^2$, **not** $(2a)^2$
$a \times a \times a = a^3$, **not** $3a$
$a \times b \times 2 = 2ab$

Collecting like terms

Expressions can be simplified by collecting **like terms**. Only collect the terms if their letters and powers are **identical**. For example:

$4a + 2a = 6a$

$3a^2 + 6a^2 - 4a^2 = 5a^2$

$4a + 6b - 3a + 2b = a + 8b$ — Add the a terms together, then the terms with b. Remember a means $1a$.

$9a + 4b$ — Cannot be simplified since there are no like terms.

$3xy + 2yx = 5xy$ — Remember xy means the same as yx.

The diagram shows a floor plan.

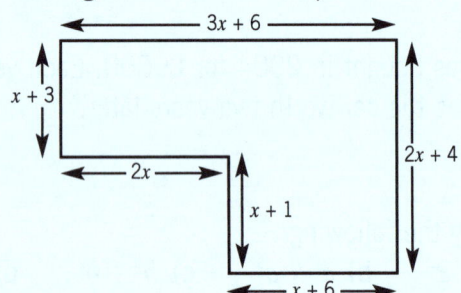

The perimeter of the floor in terms of x is
$3x + 6 + 2x + 4 + x + 6 + x + 1 + 2x + x + 3$
$= 10x + 20$

Know the words

Formula – connects two expressions containing variables, the value of one variable depending on the values of the others. It must have an equals sign, e.g. $v = u + at$. When the values of u, a and t are known, the value of v can be found using the formula.

Equation – connects two expressions, involving definite unknown values. It must have an equals sign, e.g. $x + 2 = 5$. This is only true when $x = 3$.

Multiplying out brackets

Multiplying out brackets helps to simplify algebraic expressions. The term outside the brackets multiplies each separate term inside the brackets. For example:

$3(2x + 5) = 6x + 15$ $3 \times 2x = 6x, 3 \times 5 = 15$

$a(3a - 4) = 3a^2 - 4a$

$b(2a + 3b - c) = 2ab + 3b^2 - bc$

If the term outside the bracket is **negative**, all of the signs of the terms inside the bracket are **changed** when multiplying out.

For example:

$-4(2x + 3) = -8x - 12$ To simplify expressions, first expand the brackets, then collect like terms.

$-2(4 - 3x) = -8 + 6x$

Examples

a) Expand and simplify $2(x - 3) + 3(x + 4)$

$2(x - 3) + 3(x + 4)$
$= 2x - 6 + 3x + 12$ Multiply out the brackets. Collect like terms.
$= 5x + 6$

b) Expand and simplify $6y - 2(y - 3)$

$6y - 2(y - 3)$
$= 6y - 2y + 6$ Take care when multiplying by a negative number.
$= 4y + 6$

💡 If you are asked to 'expand' brackets it just means multiply them out. When you have finished multiplying out the brackets, simplify by collecting like terms in order to gain the final mark.

Factorisation (putting brackets in)

Factorisation is the reverse of **expanding brackets**. An expression is put into brackets by taking out **common factors**. For example:

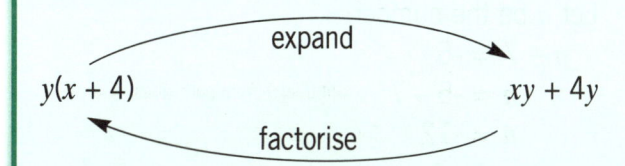

$y(x + 4)$ ⟶ expand ⟶ $xy + 4y$
         ⟵ factorise ⟵

To factorise $xy + 4y$, follow these steps:
1. Recognise that y is a factor of each term.
2. Take out this common factor.
3. The expression is completed inside the bracket, so that the result is equivalent to $xy + 4y$ when multiplied out.

Examples
Factorise the following expressions.

a) $4a + 8$ = $4(a + 2)$
b) $3b + 6$ = $3(b + 2)$
c) $3x^2 + 9$ = $3(x^2 + 3)$
d) $5x^2 + x$ = $x(5x + 1)$
e) $4m^2 + 8m$ = $4m(m + 2)$
f) $10a^2 - 15a$ = $5a(2a - 3)$
g) $12xy - 16x$ = $4x(3y - 4)$
h) $12x - 4$ = $4(3x - 1)$
i) $x^2 - 6x$ = $x(x - 6)$

Quick test

1. Simplify these expressions by collecting like terms.
   a) $5a + 2a + 3a$
   b) $6a - 3b + 4b + 2a$
   c) $5x - 3x + 7x - 2y + 6y$
   d) $3xy^2 - 2x^2y + 6x^2y - 8xy^2$

2. Multiply out the brackets and simplify where possible.
   a) $3(x + 2)$
   b) $2(x + y)$
   c) $-3(2x + 4)$
   d) $5(x + 2) + 3(x + 1)$
   e) $2(x - 1) + 5(x + 2)$
   f) $6(2x + 1) - 2(x - 3)$

3. Factorise the following:
   a) $4x^2 + 2$
   b) $2y - 4$
   c) $6x + 10$
   d) $3x^2 + 9x$
   e) $15x - 20x^2$
   f) $x^2 - 5x$

Equations 1

Equations

An **equation** involves finding an unknown value that has to be worked out.

Here is an example of an equation:
$3x - 5 = 10$

Solving simple linear equations

An equation has two parts separated by an equals sign. When working out the unknown value in an equation, the **balance method** is used; that is, whatever you do to one side of an equation you must do the same to the other side.

Examples

a) Solve these equations.

i) $n - 7 = 10$
$n = 10 + 7$ — Add 7 to both sides.
$n = 17$

ii) $n + 6 = 8$
$n = 8 - 6$ — Subtract 6 from both sides.
$n = 2$

iii) $3n = 21$
$n = \frac{21}{3}$ — Divide both sides by 3.
$n = 7$

iv) $\frac{n}{4} = 3$
$n = 3 \times 4$ — Multiply both sides by 4.
$n = 12$

v) $\frac{15}{n} = 5$
$15 = 5n$ — Multiply both sides by n.
$\frac{15}{5} = n$ — Divide both sides by 5.
$n = 3$

b) I think of a number, add 7 and my answer is -5. What is my number?

Let n be the number:
$n + 7 = -5$
$n = -5 - 7$ — Subtract 7 from both sides.
$n = -12$

Solving linear equations of the form $ax + b = c$

Remember, use the balance method; whatever is done to one side of the equation must be done to the other.

Examples

a) Solve $2x + 15 = 9$
$2x = 9 - 15$ — Subtract 15 from both sides.
$2x = -6$
$x = -6 \div 2$ — Divide both sides by 2.
$x = -3$

b) Solve $\frac{n}{3} + 2 = 6$
$\frac{n}{3} = 6 - 2$ — Subtract 2 from both sides.
$\frac{n}{3} = 4$
$n = 4 \times 3$ — Multiply both sides by 3.
$n = 12$

c) Solve $3x - 1 = 7$
$3x = 7 + 1$
$3x = 8$
$x = \frac{8}{3} = 2\frac{2}{3}$

d) Solve $\frac{n}{4} - 2 = 4$
$\frac{n}{4} = 4 + 2$
$\frac{n}{4} = 6$
$n = 6 \times 4$
$n = 24$

Solving linear equations of the form $ax + b = cx + d$

The trick with this type of equation is to get the x terms together on one side of the equals sign and the numbers on the other side.

Examples

a) Solve $6x - 4 = 4x + 8$

$$6x - 4 - 4x = 8 \quad \text{← Subtract } 4x \text{ from both sides.}$$
$$6x - 4x = 8 + 4 \quad \text{← Add 4 to both sides.}$$
$$2x = 12$$
$$x = \frac{12}{2} \quad \text{← Divide each side by 2.}$$
$$x = 6$$

b) Solve $5x - 9 = 12 - 4x$

$$9x - 9 = 12 \quad \text{← Add } 4x \text{ to both sides.}$$
$$9x = 21 \quad \text{← Add 9 to both sides.}$$
$$x = \frac{21}{9} = 2\tfrac{1}{3}$$

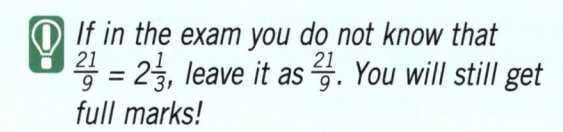

💡 *If in the exam you do not know that $\frac{21}{9} = 2\tfrac{1}{3}$, leave it as $\frac{21}{9}$. You will still get full marks!*

Solving linear equations with brackets

Just because an equation has brackets, do not be put off. It is just the same as the other equations once the brackets have been multiplied out.

Examples

a) Solve $3(x - 4) = 21$

$$3x - 12 = 21$$
$$3x = 21 + 12$$
$$3x = 33$$
$$x = \frac{33}{3}$$
$$x = 11$$

Multiply out the brackets first. Then solve as before.

b) Solve $3(x - 2) = 2(x + 6)$

$$3x - 6 = 2x + 12$$
$$3x - 6 - 2x = 12$$
$$x - 6 = 12$$
$$x = 12 + 6$$
$$x = 18$$

c) Solve $5(n - 2) + 6 = 3(n - 4) + 10$

$$5n - 10 + 6 = 3n - 12 + 10$$
$$5n - 4 = 3n - 2$$
$$2n = 2$$
$$n = 1$$

Check: $5(1 - 2) + 6 = 3(1 - 4) + 10$
$$-5 + 6 = -9 + 10$$
$$1 = 1 ✔$$

💡 *Solving equations is a very common topic at GCSE. Try to work through them in a logical way, always showing full working out. If you have time, check your answer by substituting it back into the original equation to see if it works.*

Quick test

1. Solve the following equations.

   a) $x + 6 = 10$
   b) $2x = 12$
   c) $\frac{x}{5} = 4$
   d) $x - 4 = 9$
   e) $\frac{x}{2} - 3 = 9$
   f) $4x + 2 = 20$
   g) $5x + 3 = 2x + 9$
   h) $6x - 1 = 15 + 2x$
   i) $3(x + 2) = x + 4$
   j) $2(x - 1) = 6(2x + 2)$

Equations 2

Using equations to solve problems

Examples

a) The perimeter of the triangle is 20cm. Work out the value of x and hence find the lengths of the three sides.

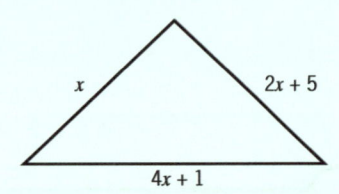

(triangle with sides x, $2x+5$, $4x+1$)

$$x + 2x + 5 + 4x + 1 = 20$$
$$7x + 6 = 20$$
$$7x = 20 - 6$$
$$7x = 14$$
$$x = \frac{14}{7}$$
$$x = 2$$

The perimeter is found by adding the lengths together. Collect like terms and solve the equation as before.

So the lengths of the sides are 2 (= x), 9 (= $4x + 1$) and 9 (= $2x + 5$).

b) Rhysian works for y hours each week for 2 weeks. In the third week, she works for 3 hours longer and in the fourth week she works for 7 hours longer. She works a total of 154 hours during these 4 weeks. By forming an equation, work out…

i) the value of y

$$y + y + y + 3 + y + 7 = 154$$ *Write an equation.*
$$4y + 10 = 154$$
$$4y = 154 - 10$$
$$4y = 144$$
$$y = \frac{144}{4}$$
$$y = 36 \text{ hours}$$

ii) the number of hours Rhysian works in the third week.

During the third week Rhysian works $y + 3$ hours, so she works $36 + 3 = 39$ hours.

Simultaneous equations

Two equations with two unknowns are called **simultaneous equations**. They can be solved in several ways. You only need to know how to solve them using a graphical method. Solving equations simultaneously involves finding values for the letters that will make both equations work.

Graphical method

The points at which any two graphs intersect represent the simultaneous solutions of their equations.

Example

Solve the simultaneous equations $y = 2x - 1$, $x + y = 5$

- First draw the two graphs.

  $y = 2x - 1$ If $x = 0$, $y = -1$
  If $y = 0$, $x = \frac{1}{2}$
  If $y = 1$, $x = 1$

  $x + y = 5$ If $x = 0$, $y = 5$
  If $x = 5$, $y = 0$

- This is the solution. At the point of intersection, $x = 2$ and $y = 3$

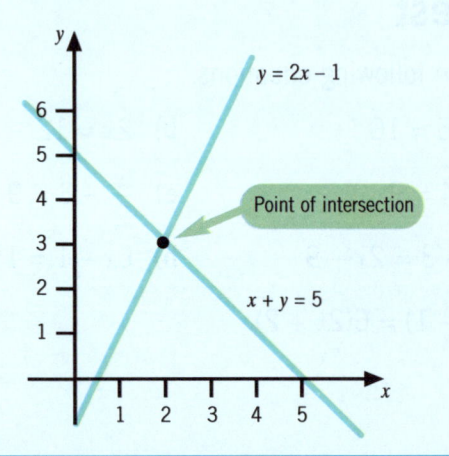

(graph showing lines $y = 2x - 1$ and $x + y = 5$ intersecting at point of intersection)

Solving cubic equations by trial and improvement

Trial and improvement is a method of solving equations by substituting values until a close value is found. This method is particularly useful when solving **cubic equations**.

Example
The equation $x^3 - 5x = 10$ has a solution between 2 and 3. Find this solution to 2 decimal places.

Draw a table to help.
Substitute different values of x into $x^3 - 5x$.

x	$x^3 - 5x$	Comment
2.5	3.125	Too small
2.8	7.952	Too small
2.9	9.889	Too small
2.95	10.922375	Too big
2.94	10.712184	Too big
2.91	10.092171	Too big

At this stage we know the solution is somewhere between 2.90 and 2.91

Checking the middle value, $x = 2.905$, gives $x^3 - 5x = 9.99036...$ which is too small.

```
    |              |              |
   2.90          2.905          2.91
(too small)   (too small)    (too big)
```

The diagram shows that the solution is between 2.905 and 2.91, so $x = 2.91$ correct to 2 decimal places.

💡 *Make sure you write down the solution for x, not the answer to $x^3 - 5x$.*

Quick test

1. The diagram shows the graphs of the equations $x + y = 2$ and $y = x - 4$
   Use the diagram to solve the simultaneous equations $x + y = 2$ and $y = x - 4$

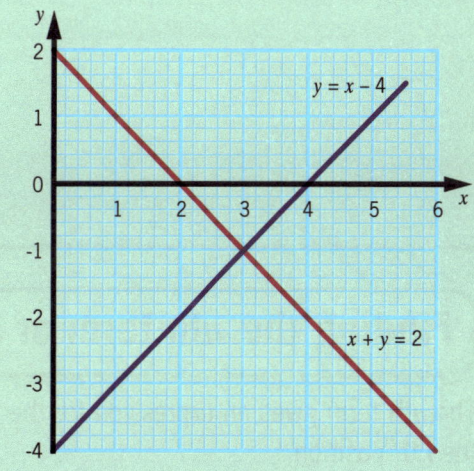

2. The equation $y^3 + y = 40$ has a solution between 3 and 4. Find this solution to 1 decimal place by using the method of trial and improvement.

3. The width of a rectangle is x centimetres.
   The length of the rectangle is $(x + 2)$ centimetres.
   a) Find an expression in terms of x for the perimeter of the rectangle.
      Give your expression in its simplest form.
   b) The perimeter of the rectangle is 40cm.
      Work out the length of the rectangle.

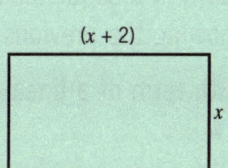

Patterns, sequences & inequalities

Sequences

A **sequence** is a list of numbers. There is usually a relationship between the numbers. Each value in the list is called a **term**.

For example, the odd numbers form a sequence 1, 3, 5, 7, 9, 11, ..., in which the terms have a common difference of 2.

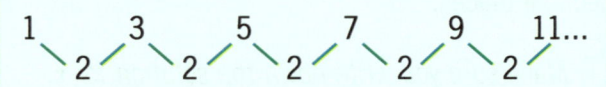

1 →(2) 3 →(2) 5 →(2) 7 →(2) 9 →(2) 11...

Important number sequences
Even numbers: 2, 4, 6, 8, 10, ...
Square numbers: 1, 4, 9, 16, 25, ...
Cube numbers: 1, 8, 27, 64, 125, ...
Powers of 2: 1, 2, 4, 8, 16, ...
Triangular numbers: 1, 3, 6, 10, 15, ...
Powers of 10: 1, 10, 100, 1000, ...

Number patterns

When a number pattern is continued, the numbers relating to a pattern often follow a rule.

Example
Here are the first three patterns in a sequence of patterns made from matchsticks:

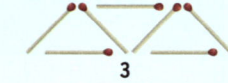

a) Draw pattern number 4.

b) Complete the table:

Pattern number	1	2	3	4	5
Number of matchsticks	3	5	7	9	11

c) Explain how you would work out pattern number 7.

Keep adding 2 to the number of matchsticks **or** multiply the pattern number by 2 and add 1.

d) Work out how many matchsticks are in pattern number 13.

$13 \times 2 + 1 = 27$ matches

Finding the nth term of a linear sequence

The nth term gives an expression for the term in the nth position.

The nth term of a sequence is useful as it allows any term in a sequence to be found without relying on knowing the previous term.

The nth term of a linear sequence is of the form $an + b$.

Examples
a) Find an expression for the nth term of this sequence: 3, 5, 7, 9, ...

① Find the common difference; this is a.
Difference = 2, so $a = 2$.
So the nth term = $2n + b$

Term	1	2	3	4 .... n
Number in sequence	3	5	7	9 ....

② Now substitute the value of $n = 1$ and the number in the sequence; in this case it is 3.
nth term = $2n + b$
$3 = 2 \times 1 + b$
$3 = 2 + b$
$3 - 2 = b$, so $b = 1$
nth term is $2n + 1$

③ Check when $n = 2$: $2 \times 2 + 1 = 5$
Number in sequence is 5 so the nth term is $2n + 1$.

Finding the nth term of a linear sequence (cont.)

b) In a fairground game, a prize is won if the number you pick out of a jar is in the sequence whose nth term is $3n + 1$. Josie picks a card with the number 600. Explain whether Josie wins a prize.

$3n + 1 = 600$
So, $3n = 600 - 1 = 599$
$n = \dfrac{599}{3}$

Since 599 is not divisible by 3, Josie does not win a prize.

Inequalities

The inequality symbols are as follows:
- $>$ means '**greater than**'
- $<$ means '**less than**'
- $\geqslant$ means '**greater than or equal to**'
- $\leqslant$ means '**less than or equal to**'

So $x > 3$ and $3 < x$ both say 'x is greater than 3'.

Inequalities are solved in a similar way to equations. Multiplying or dividing by **negative numbers** changes the **direction** of the sign.

For example, if $-x \geqslant 5$ then $x \leqslant -5$.

Examples
Solve the following inequalities.

a) $x + 4 < 10$

$x < 10 - 4$ — Subtract 4 from both sides.
$x < 6$

The solution of the inequality $x < 6$ can be represented on a number line:

Use ○ when the end point is not included.

b) $4x - 2 \leqslant 2x + 6$

$2x - 2 \leqslant 6$ — Subtract $2x$ from both sides.
$2x \leqslant 8$ — Add 2 to both sides.
$x \leqslant 4$ — Divide both sides by 2.

The solution of the inequality $x \leqslant 4$ can be represented on a number line:

Use • when the end point is included.

c) $-5 < 3x + 1 \leqslant 13$

$-6 < 3x \leqslant 12$ — Subtract 1 from each part.
$-2 < x \leqslant 4$ — Divide each part by 3.

The **integer values** that satisfy the above inequality are **-1, 0, 1, 2, 3, 4**.

The inequality $-2 < x \leqslant 4$ is shown on the number line below.

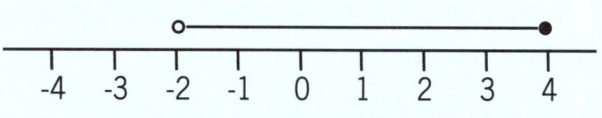

Quick test

1. Write down the next two terms in each of these sequences.
   a) 5, 7, 9, 11, ___ , ___   **b)** 1, 4, 9, 16, ___ , ___   **c)** 12, 10, 8, 6, ___ , ___

2. Write down the nth term of each of these sequences.
   a) 5, 7, 9, 11, ...   **b)** 2, 5, 8, 11, ...   **c)** 6, 10, 14, 18, ...   **d)** 8, 6, 4, 2, ...

3. Solve the following inequalities.
   a) $2x - 3 < 9$ **b)** $5x + 1 \geqslant 21$ **c)** $1 \leqslant 3x - 2 \leqslant 7$ **d)** $1 \leqslant 5x + 2 < 12$

Formulae

Writing formulae

You need to be able to write a formula when given some information or a diagram.

Examples

a) Frances buys x books at £2.50 each. She pays with a £20 note. If she receives C pounds change, write down a formula for C.

$C = 20 - 2.50x$

Notice that no £ signs are put in the formula.

This is the amount of money she spent.

If in doubt, check by substituting a value for x into the equation above. For example, if she bought 2 books $x = 2$, so her change would be $20 - 2.50 \times 2 = £15.00$ ✔

b) Some patterns are made by using grey and white paving slabs.

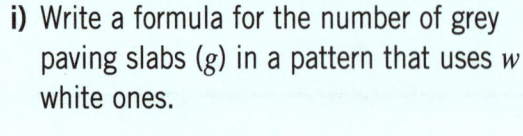

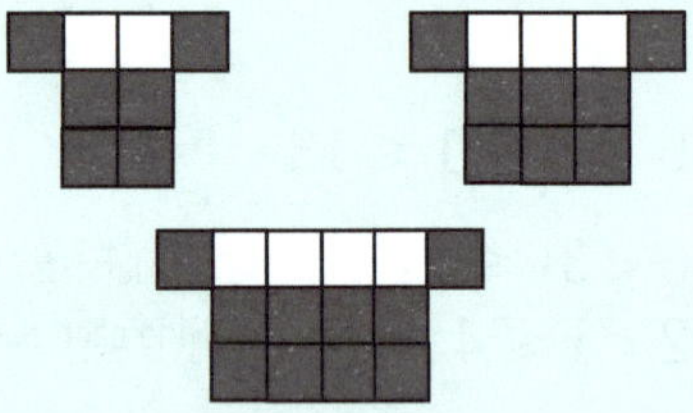

i) Write a formula for the number of grey paving slabs (g) in a pattern that uses w white ones.

The formula is $g = 2w + 2$

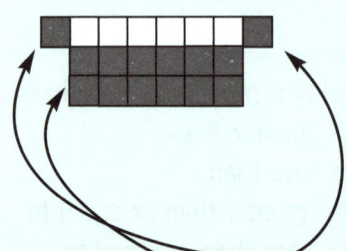

$2w$ represents the 2 layers; $+ 2$ gives the grey slabs on either end of the white ones.

ii) Andy builds a patio using the pattern shown above. He uses 24 white slabs. How many grey slabs will he need to buy?

$g = 2w + 2$
$g = 2 \times 24 + 2$
$g = 48 + 2$
$g = 50$

Andy needs to buy 50 grey slabs.

Using formulae

A formula describes the relationship between two (or more) variables.

A formula must have an equals sign (=) in it.

Examples

a) Josh hires a van. There is a standing charge of £17 and then it costs £21 per day. How much does it cost for...

i) 6 days?

$17 + (21 \times 6) = £143$

ii) y days?

$17 + (21 \times y) = £(17 + 21y)$

iii) Write a formula for the total hire cost C.

$C = 17 + 21y$

This is a formula that works out the cost of hiring the van for any number of days.

b) A rule for working out the average carbon dioxide emission from a coach is given by:

| Carbon dioxide emission | = | Distance travelled (km) × 0.17 |

If C stands for the carbon dioxide emission in grams and d stands for the distance travelled in kilometres...

i) write down a formula for the total carbon dioxide emission, C

$C = d \times 0.17$ or $C = 0.17d$

ii) work out the carbon dioxide emission for a bus that travels a distance of 28km.

$C = 28 \times 0.17$
$C = 4.76$ grams

Formulae, expressions and substituting

$p + 3$ is an **expression**.
$y = p + 3$ is a **formula**. The value of y depends on the value of p.

Replacing a letter with a number is called **substitution**. When substituting...
- write out the expression first and then replace the letters with the values given
- work out the value on your calculator. Use bracket keys where necessary and pay attention to the order of operations (BIDMAS).

Examples
a) Using $a = 6$, $b = -3$ and $c = 5$, find the values of these expressions.

i) $5a + 2c$

$5a + 2c = 5 \times 6 + 2 \times 5$
$= 30 + 10$
$= 40$

ii) $\dfrac{3a}{b}$

$\dfrac{3a}{b} = \dfrac{3 \times 6}{-3} = -6$

iii) $2c^2 + 4$

$2c^2 + 4 = 2 \times 5^2 + 4 = 50 + 4 = 54$

iv) $3a^2$

$3a^2 = 3 \times 6^2 = 3 \times 36 = 108$

💡 *Remember to show the substitution.*

b) The formula $C = \dfrac{5(F - 32)}{9}$ is used to change the temperature from degrees Fahrenheit (F) into degrees centigrade (C). Work out the value of C when $F = -12$.

$C = \dfrac{5(-12 - 32)}{9}$ — Substitute the value of F into the formula.

$C = \dfrac{5 \times (-44)}{9}$ — Work out the brackets first, then multiply by $\dfrac{5}{9}$

$C = -24.\dot{4}°C$ or $-24\dfrac{4}{9}°C$

💡 *When substituting into an expression or formula you must show each step in your working out. By showing your substitution you will obtain method marks even if you get the final answer wrong.*

Rearranging formulae

The **subject of a formula** is the letter that appears on its own on one side of the formula.

Examples
Make a the subject of these formulae.

a) $v = u + at$

$v - u = at$ — Subtract u from both sides.
$\dfrac{v - u}{t} = a$ — Divide both sides by t.
So $a = \dfrac{v - u}{t}$ — The subject of the formula is usually written first.

b) $b = 3(a - 2)$

$\dfrac{b}{3} = a - 2$ — Divide both sides by 3.
$\dfrac{b}{3} + 2 = a$ — Add 2 to both sides.
So $a = \dfrac{b}{3} + 2$

c) $x = ay$

$\dfrac{x}{y} = a$ — Divide both sides by y.
So $a = \dfrac{x}{y}$

Quick test

1 Using $p = 4$, $q = -2$ and $r = 3$, find the value of these expressions.

a) $3p - 2$ b) $r^2 + 5$ c) $q^2 + 4$ d) $\dfrac{2r^2}{3}$ e) $7q - 4r$

2 Make b the subject of the formula $a = 5b - d$

3 Make m the subject of the formula $y = mx + c$

4 Make R the subject of the formula $V = IR$

Straight-line graphs

Coordinates

- **Coordinates** are used to locate the position of a point.
- When reading coordinates, read **across first, then up or down**.
- Coordinates are always written in **brackets**, and with a **comma** in between, e.g. (2, 4).
- The horizontal axis is the *x*-axis. The vertical axis is the *y*-axis.

On the grid opposite...
- A has coordinates (2, 4)
- B has coordinates (-1, 3)
- C has coordinates (-2, -3)
- D has coordinates (3, -1)

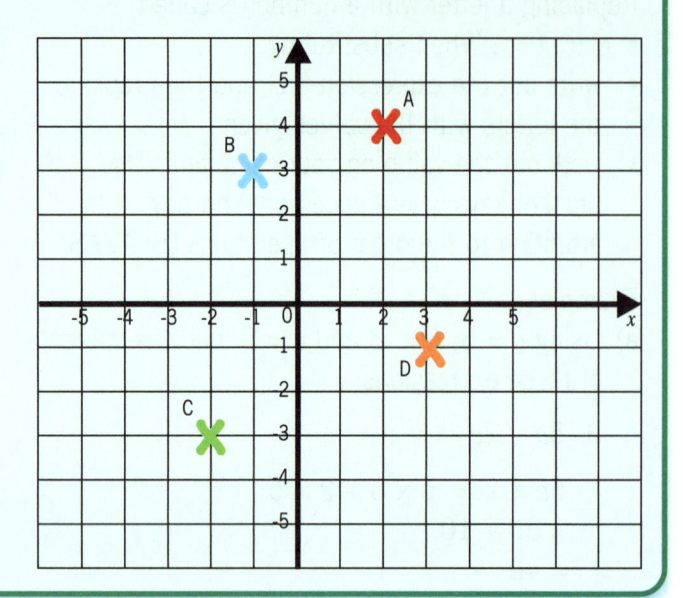

Drawing straight-line graphs

To draw a straight-line graph, follow these easy steps:

1. Choose three values of *x* and draw up a table.
2. Work out the value of *y* for each value of *x*.
3. Plot the coordinates and join up the points with a straight line.
4. Label the graph.

Example
Draw the graph of $y = 3x - 1$

1. Draw a table with some suitable values of *x*.
2. Work out the *y* values by substituting each *x* value into the equation.
   e.g. $x = -2 \rightarrow y = 3 \times -2 - 1$
   $= -6 - 1 = -7$

x	-2	0	2
$y = 3x - 1$	-7	-1	5

3. Plot the points and draw the line.

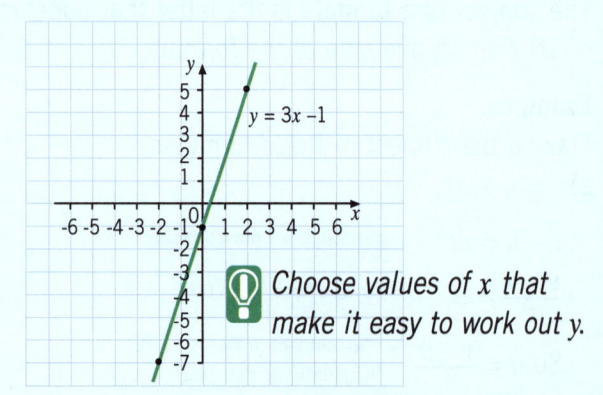

Choose values of x that make it easy to work out y.

Graphs of $y = a$, $x = b$, $x + y = k$ and $x = y$

$y = a$ is a **horizontal** line with every *y* coordinate equal to *a*.

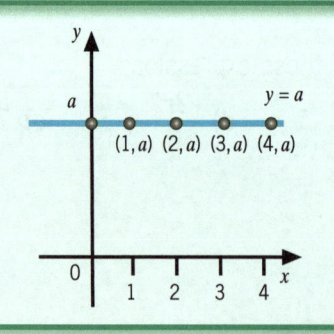

$x = b$ is a **vertical** line with every *x* coordinate equal to *b*.

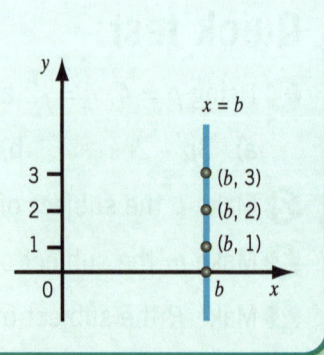

Graphs of $y = a$, $x = b$, $x + y = k$ and $x = y$ (cont.)

Graphs of $x + y = k$, where k is a number, always give straight-line graphs. Tables of values are not always needed when drawing this type of graph.

Example
Draw the graph of $x + y = 3$

On the x-axis, $y = 0$ so $x = 3$ since $3 + 0 = 3$
On the y-axis, $x = 0$ so $y = 3$ since $0 + 3 = 3$
The graph goes straight through the points (3, 0) and (0, 3).

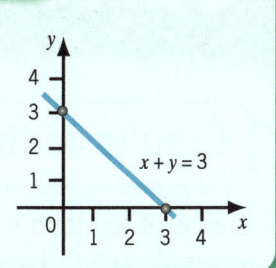

Interpreting $y = mx + c$

The general equation of a straight-line graph is

$$y = mx + c$$

where m is the **gradient** (steepness) of the line.

Remember these points about straight-line graphs:
- As m **increases**, the line gets **steeper**.
- If m is **positive**, the line slopes **forwards**.
- If m is **negative**, the line slopes **backwards**.
- c is the **intercept** on the y-axis, that is, where the graph cuts the y-axis.
- **Parallel** lines have the **same gradient**.

💡 It will help if you can sketch a straight-line graph from its equation. If you can do this, you will be able to tell if the graph you have drawn is correct.

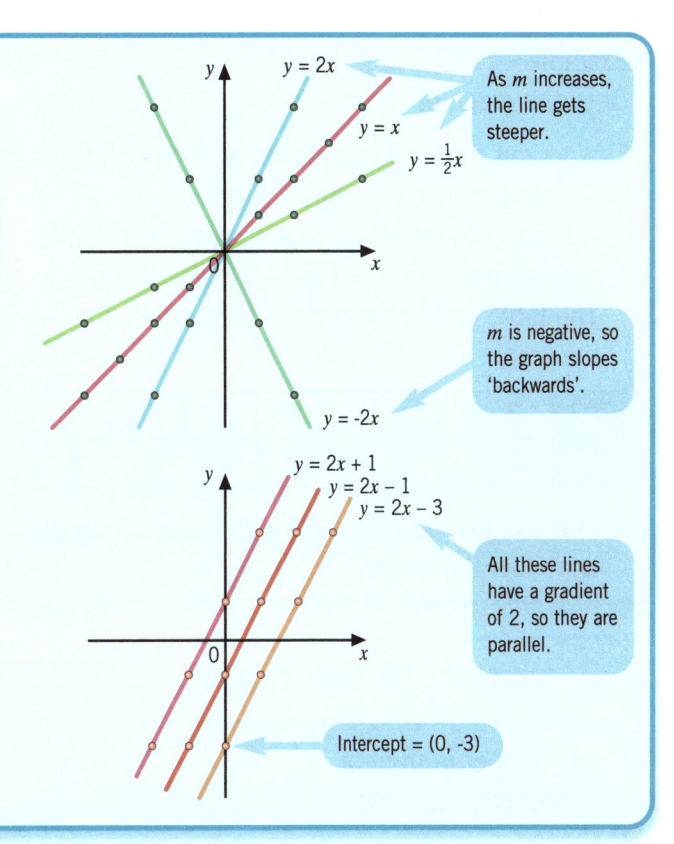

As m increases, the line gets steeper.

m is negative, so the graph slopes 'backwards'.

All these lines have a gradient of 2, so they are parallel.

Intercept = (0, -3)

Finding the gradient of a line

To find the **gradient** of a line, follow these steps:
1. Choose two points on the line.
2. Draw a triangle as shown.
3. Find the change in y (height) and the change in x (base).
4. Gradient = $\dfrac{\text{change in } y}{\text{change in } x}$ or $\dfrac{\text{height}}{\text{base}} = \dfrac{4}{3} = 1\dfrac{1}{3}$
5. Decide if the gradient is positive or negative (see above).

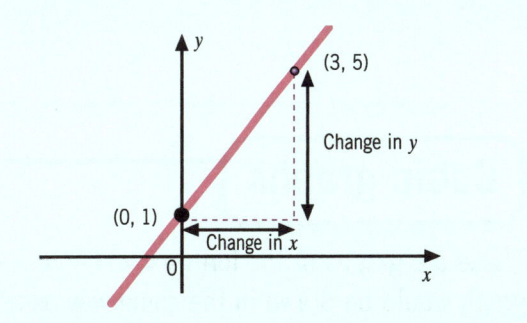

💡 Do not count the squares as the scales may be different.

Quick test

1. Draw the graph of $y = 6 - 2x$. From your graph, write down the solution of the following equations.
   a) $6 - 2x = 4$ b) $6 - 2x = 3$

2. Which graph is steeper?
   $y = 4x - 1$ $y = 3x + 2$

Curved graphs

Graphs of the form $y = ax^2 + b$

Graphs of the form $y = ax^2 + b$ are known as **quadratic graphs** because they have a square term in them. They are sometimes called **parabolas**.

Example
Draw the graph of $y = x^2 - 3$.

❶ Work out the y coordinates for several points. Remember that x^2 means x times x.
Replace x in the equation with each coordinate, e.g. when $x = -3$, $y = (-3)^2 - 3 = 9 - 3 = 6$
The table gives coordinates of the points for the graph.

x	-3	-2	-1	0	1	2	3
y	6	1	-2	-3	-2	1	6

❷ Plot and join up the points with a smooth curve, and label the graph.

If you are asked to draw the graph of $y = 2x^2$, remember this means $y = 2 \times (x^2)$. (Square x first, then multiply by 2.)

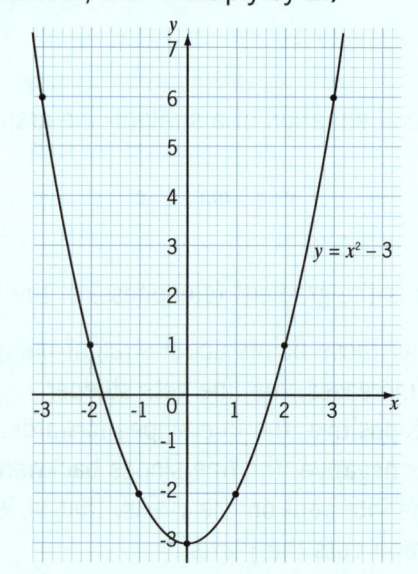

Direction of the curve

If the number in front of x^2 is positive, the curve is U-shaped.

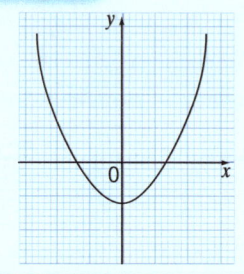

If the number in front of x^2 is negative, the curve is an upside-down U.

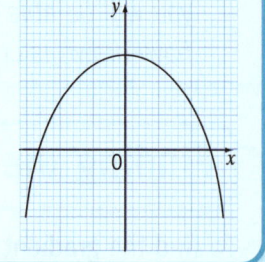

Cubic graphs

These are graphs of the form $y = ax^3 + b$. A cubic graph would be drawn in the same way as a quadratic graph.

Example
Draw the graph of $y = x^3 - 2$

x	-2	-1	0	1	2
y	-10	-3	-2	-1	6

Remember x^3 means $x \times x \times x$.

Work out the table of values that give the coordinates of the points on the graph.

Plot the points and join with a smooth curve.

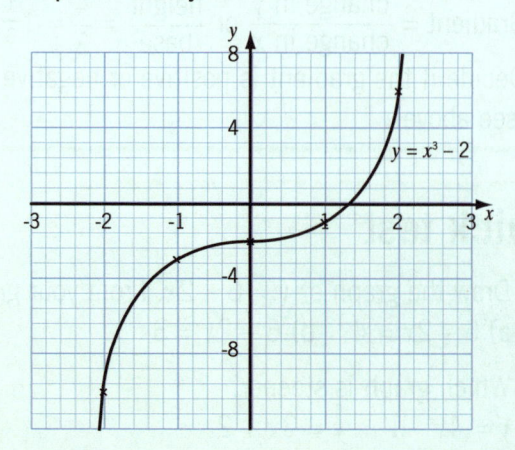

Graphs of the form $y = ax^2 + bx + c$

The following example will help you to understand graphs of the form $y = ax^2 + bx + c$.

Example
Draw the graph of $y = x^2 - x - 6$ using values of x from -2 to 3. Use the graph to find the value of x when $y = -3$.

① Make a table of values.
② Work out the values of y by substituting the values of x into the equation.
For example, if $x = 1$, $y = x^2 - x - 6$
   $= 1^2 - 1 - 6$
   $= -6$

Do not try to input this all into your calculator at once. Do it step by step.

x	-2	-1	0	1	2	3	0.5
y	0	-4	-6	-6	-4	0	-6.25

$x = 0.5$ is worked out to find the minimum value.

③ Plot the points and join them with a smooth curve.

From the curve below it is clear that...
- the **line of symmetry** is at $x = 0.5$
- the **minimum value** is when $x = 0.5$, $y = -6.25$
- the curve cuts the y-axis at (0, -6), i.e. (0, c).

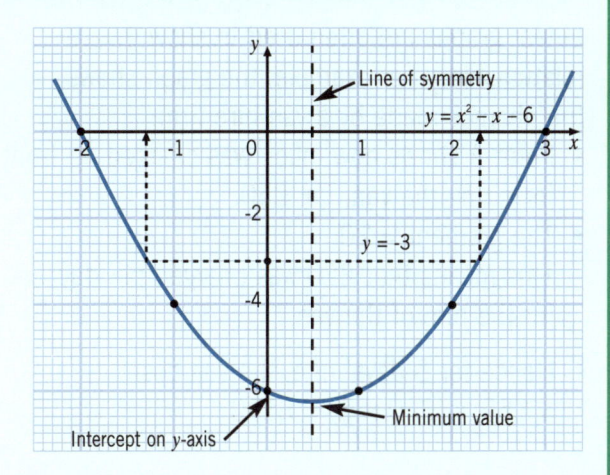

Show clearly on your graph how you take your readings.

④ When $y = -3$, read across from $y = -3$ to the graph then read up to the x-axis: $x = -1.3$ and $x = 2.3$
These are the approximate solutions of the equation $x^2 - x - 6 = -3$.

Draw the curve with a sharp pencil. Go through all the points and check for any parts that look wrong.

Quick test

① a) Complete the table of values for the graph $y = x^2 + 3$.

x	-3	-2	-1	0	1	2	3
y							

b) Draw the graph of $y = x^2 + 3$.
c) From the graph, find the value of x when $y = 8$.

② Complete the table of values and draw the graph of $y = 2x^2 - 1$ for values of x from -3 to 3.

x	-3	-2	-1	0	1	2	3
y							

Algebra

43

Interpreting graphs

Graphs in practical situations

Linear graphs are often used to show **relationships**. For example, the graph shows the charges made by a van-hire firm:
- Point A shows how much was charged up front for hiring the van, i.e. £50.
- The gradient = 20.
  This means that £20 was charged per day for the hire of the van.

  Hence for 5 days' hire, the van cost
  = £50 + (£20 × 5) = £150

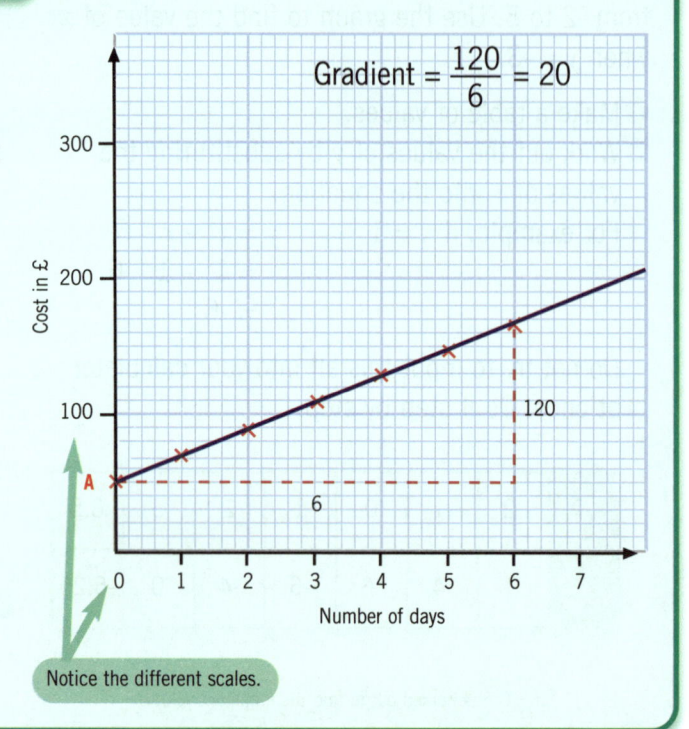

Notice the different scales.

Distance–time graphs

Distance–time graphs are often called **travel graphs**. The **speed** of an object can be found by finding the gradient of the line:

$$\text{Speed} = \frac{\text{Distance travelled}}{\text{Time taken}}$$

Example
The graph represents Mr Rogers' car journey; he sets off at 9am (0900). Work out the speed of each stage of the journey.

a) The car is travelling at 30mph for 1 hour (30 ÷ 1).
b) The car is stationary for 30 minutes.
c) The graph is steeper so the car is travelling faster, at a speed of 60mph for 30 minutes (30 ÷ 0.5).
d) The car is stationary for 1 hour.
e) The return journey is at a speed of 40mph (60 ÷ 1.5).

Notice the importance of using the gradient of a line. It is useful to note that on this distance–time graph example, the scales on the axes are different. Care must be taken when reading the scales: always make sure you understand the scales before you start.

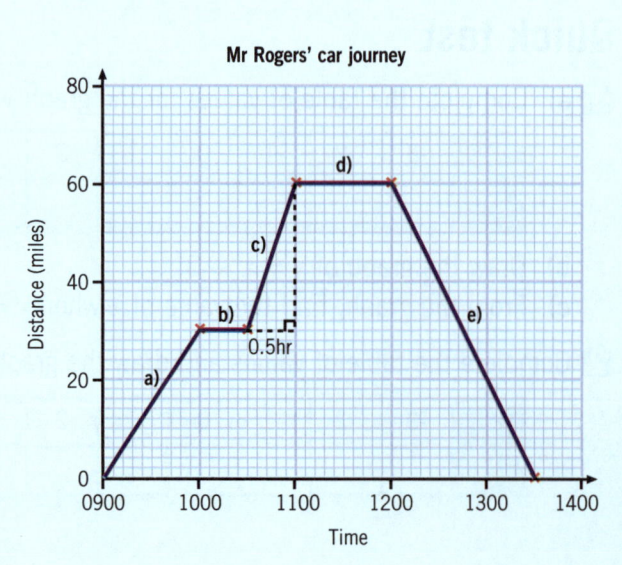

Conversion graphs

Conversion graphs are used to convert values of one quantity to another, e.g. litres to pints, kilometres to miles, pounds to dollars, etc.

Example
Suppose £1 is worth $1.50. Draw a conversion graph.

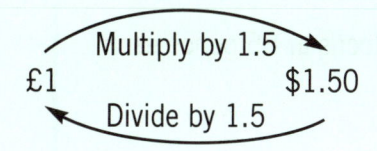

£1 — Multiply by 1.5 → $1.50
£1 ← Divide by 1.5 — $1.50

❶ Make a table of values.

£	1	2	3	4	5
$	1.5	3.0	4.5	6.0	7.5

× 1.5

❷ Plot each of these points on the graph paper.

To change $ to £, read across to the line then down. For example, $4 is £2.65 (approx.).
To change £ to $, read up to the line then across. For example, £4.50 is $6.80 (approx.).

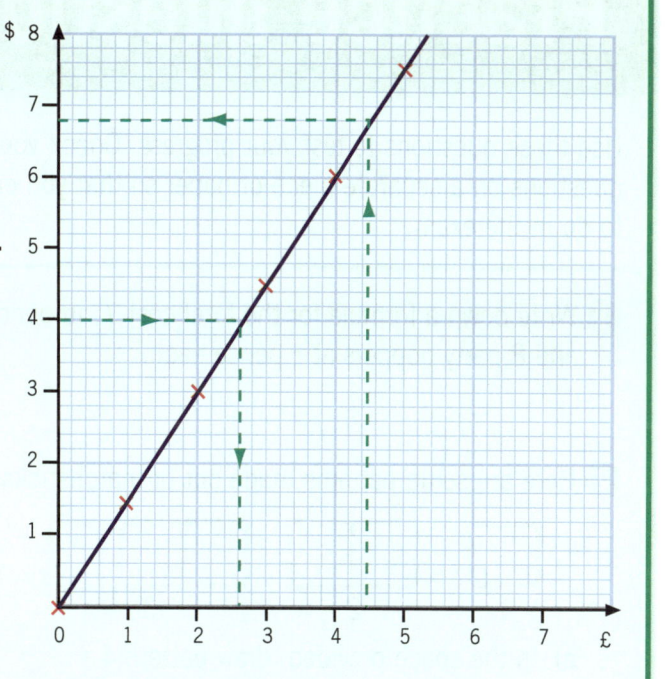

Quick test

❶ Water is being poured into these containers at a rate of 250ml per second. The graphs below show how the height of the water changes with time. Match the containers with the graphs.

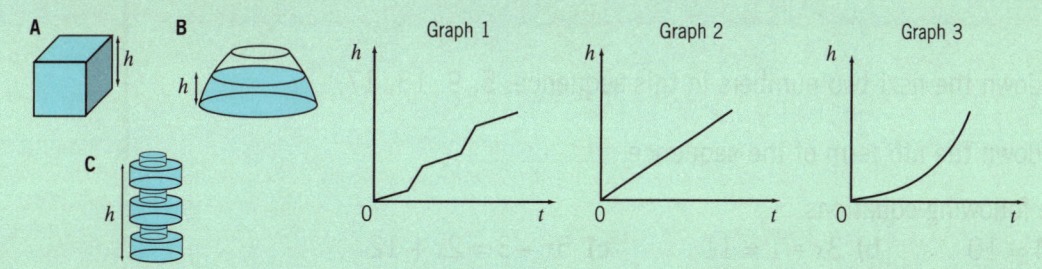

A, B, C (containers)
Graph 1, Graph 2, Graph 3

❷ The travel graph shows the car journeys of two people. From the travel graph find…

a) the speed at which Miss Young is travelling
b) the length of time Mr Price has a break
c) the speed of Mr Price from London to Birmingham
d) the time at which Miss Young and Mr Price pass each other.

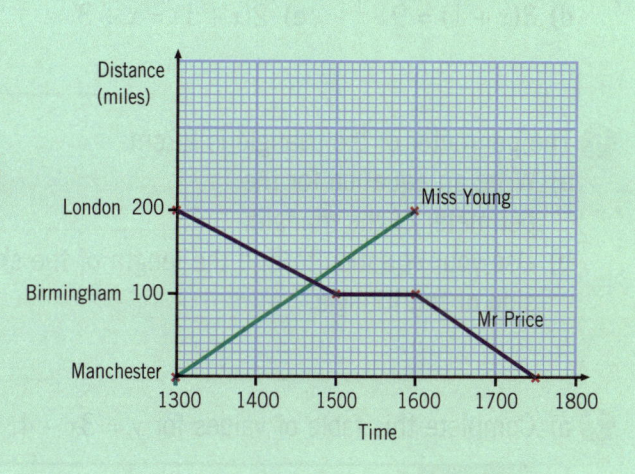

Algebra

45

Practice questions

Use these questions to test your progress. Check your answers on page 95. You may wish to answer these questions on a separate piece of paper so that you can show full working out, which you will be expected to do in the exam.

1 Write down a formula for the total cost T, in pence, of y balloons at 85 pence each and 8 party poppers at z pence each.

2 Here are some patterns made out of matchsticks:

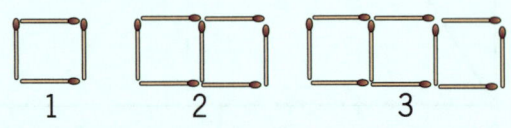

a) In the space provided, draw pattern 4.

The table shows the number of matchsticks needed for pattern 1 to pattern 3.

b) Complete the table.

c) How many matchsticks are needed for pattern 100? _____

d) Write down a formula connecting the pattern number (p) and the number of matchsticks (m). _____

Pattern number (p)	Number of matchsticks (m)
1	4
2	7
3	10
4	
5	
6	

3 Multiply out these brackets, and simplify where necessary.
a) $5(n + 2)$ b) $3(n - 2)$ c) $2(n - 5)$ d) $-3(n - 6)$ e) $5(n - 2) + 3(n - 4)$

4 a) Write down the next two numbers in this sequence: 5, 9, 13, 17, ____, ____

b) Write down the nth term of the sequence _____

5 Solve the following equations.
a) $2x + 4 = 10$ b) $3x - 1 = 11$ c) $5x - 3 = 2x + 12$
d) $3(x + 1) = 9$ e) $2(x + 1) = x + 3$

6 The perimeter of the triangle is 22cm.
a) Write an equation for the perimeter of the triangle. _____
b) Use your equation to find the length of the shortest side of the triangle.

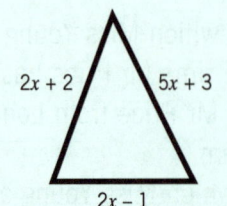

(Triangle with sides $2x + 2$, $5x + 3$, $2x - 1$)

7 a) Complete this table of values for $y = 3x - 4$.

x	-2	-1	0	1	2	3
$y = 3x - 4$			-4			

b) On graph paper, plot your values for x and y. Join your points with a straight line.

c) Write down the coordinates of the points where your line crosses the x-axis.

8 A formula for the wage earned is: wage earned = hours worked × rate per hour. Amy works for 26 hours per week. How much wage does Amy earn if she earns £7.50 per hour?

9 Simplify these expressions by collecting like terms.
   a) $a + a + a + a$ b) $2b + 3b$ c) $5b - 2b$ d) $3c - 2c + 5c$ e) $6a + 3b - 4b + 2a$

10 Factorise the following expressions.
   a) $5a - 10$ b) $6b + 12$ c) $8y^2 + 12y$

11 Simplify:
   a) $x^5 \times x^2$ b) $6^{10} \div 6^7$ c) $3x^2 \times 4x^3$ d) $(x^4)^3$

12 If $a = b^2 - 6$
   a) calculate the value of a if $b = 3$ b) calculate the value of a if $b = -4$

13 Rearrange the formula $m = 3n + p$ to make n the subject.

14 The equation $x^3 - 2x = 2$ has a solution between 1 and 2. By using a method of trial and improvement, find this solution to 1 decimal place.

15 Solve the inequality $5n - 3 \leqslant 12$

16 Water is being poured into these containers at a rate of 250ml per second. The graphs below show how the height of the water changes with time. Match the containers with the graphs.

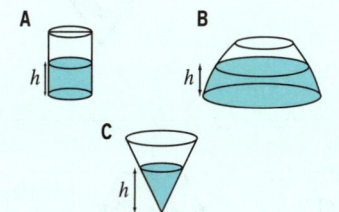

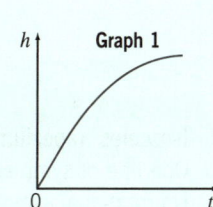

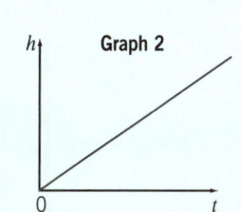

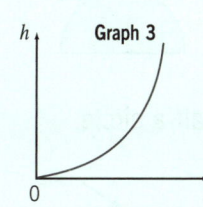

How well did you do?

| 0–6 | Try again | 7–10 | Getting there | 11–14 | Good work | 15–16 | Excellent! |

Shapes

The circle

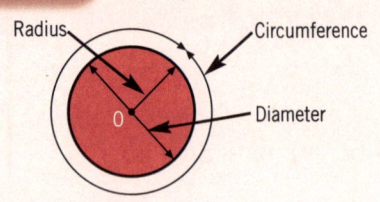

Diameter = 2 × radius
The **circumference** is the distance around the outside edge.

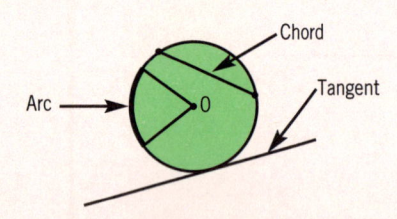

A **chord** is a line that joins two points on the circumference. A chord that goes through the centre is a diameter.
A **tangent** touches a circle at one point only.
An **arc** is part of the circumference.

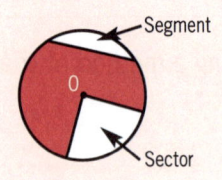

A **sector** is a part of a circle enclosed between two radii.
A **segment** is formed when chords divide a circle into different parts.

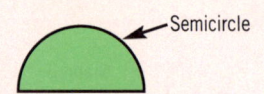

A **semicircle** is half a circle.

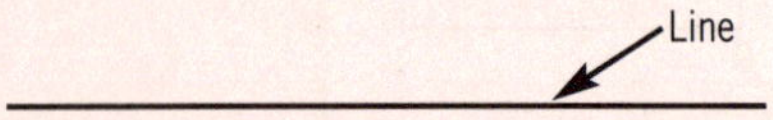

Infinitely long straight lines are just called **lines**.

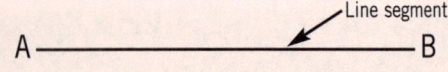

Lines that have a definite start and finish are called **line segments**.

Quadrilaterals

Quadrilaterals are four-sided shapes. You need to be able to sketch these shapes and know their symmetrical properties. See page 54 for more on symmetry.

Square
Four lines of symmetry
Rotational symmetry of order 4

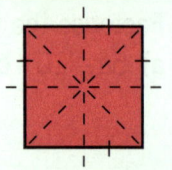

Rectangle
Two lines of symmetry
Rotational symmetry of order 2

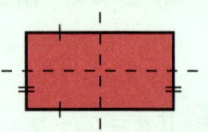

Parallelogram
No lines of symmetry
Rotational symmetry of order 2

Rhombus
Two lines of symmetry
Rotational symmetry of order 2

Kite
One line of symmetry
No rotational symmetry

Trapezium
Isosceles trapezium:
One line of symmetry
No rotational symmetry

Trapezium:
No lines of symmetry
No rotational symmetry

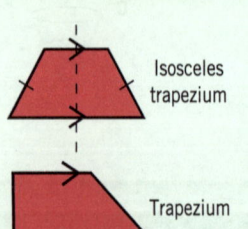

Parallel lines are lines that remain the same distance apart, i.e. they never meet.

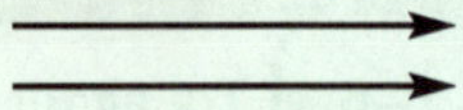

Polygons

Polygons are 2D shapes with straight sides. Regular polygons are shapes with all sides and angles equal.

💡 *Try to learn all the shapes and their symmetrical properties.*

A regular polygon can be constructed inside a circle.

Number of sides	Name of polygon
3	Triangle
4	Quadrilateral
5	Pentagon
6	Hexagon
7	Heptagon
8	Octagon
10	Decagon

Regular pentagon
It has…
- five equal sides
- a rotational symmetry of order 5
- five lines of symmetry.

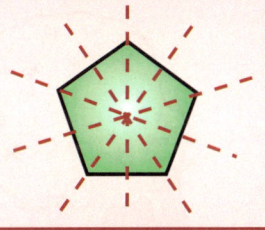

Regular hexagon
It has…
- six equal sides
- a rotational symmetry of order 6
- six lines of symmetry.

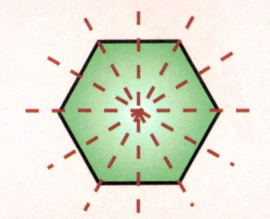

Regular octagon
It has…
- eight equal sides
- a rotational symmetry of order 8
- eight lines of symmetry.

Triangles

There are several types of triangle.

Right-angled
Has a 90° angle

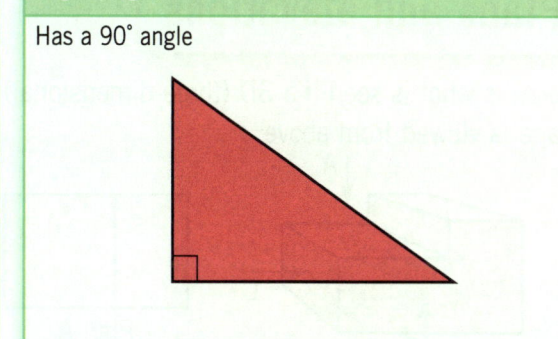

Equilateral
Three sides equal
Three angles equal
Three lines of symmetry

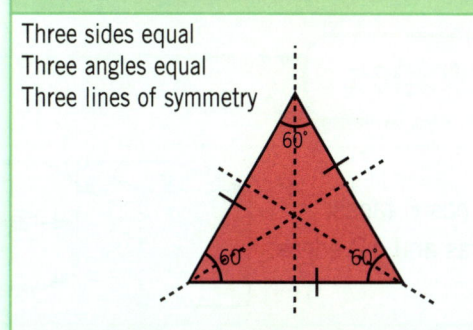

Isosceles
Two sides equal
Base angles equal
One line of symmetry

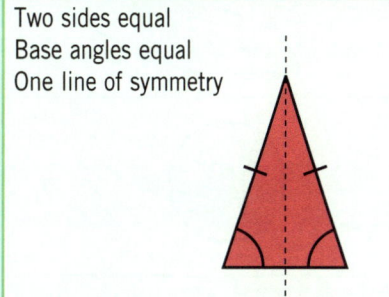

Scalene
No sides or angles the same

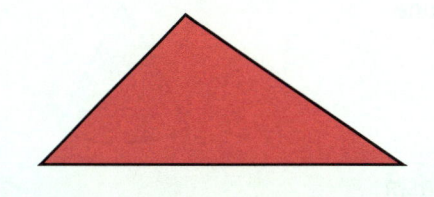

Quick test

1. What is the name of a six-sided polygon?
2. From memory, draw all the main types of triangles and quadrilaterals.

Solids

Plans and elevations

A **plan** is what is seen if a 3D (three-dimensional) shape is viewed from above.

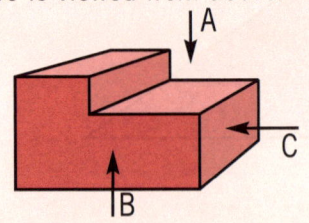

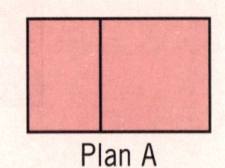

An **elevation** is seen if the 3D shape is viewed from the side or front.

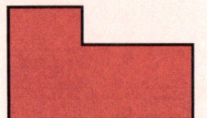

Plan A

Front elevation B

Side elevation C

3D shapes

Cube	
A cube has 6 faces, 8 vertices and 12 edges.	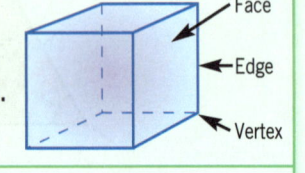 Face, Edge, Vertex
Cuboid	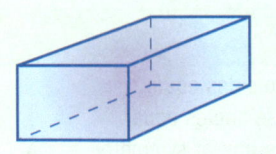
Sphere	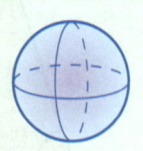
Cylinder	
Cone	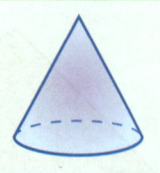
Prism A prism is a solid that can be cut into slices which are all the same shape.	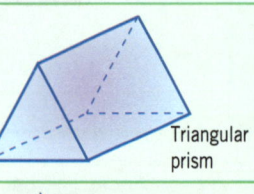 Triangular prism
Square-based pyramid	

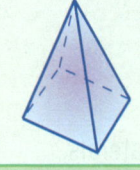

! *Remember to learn the mathematical names of the solids.*

3D shapes are seen regularly in everyday life.

Isometric drawings

You can represent 3D shapes on **isometric paper**. On this paper you can draw lengths in three perpendicular directions on the same scale. The faces do not appear in their true shapes.

A 'T'-shaped prism can be shown clearly on isometric paper (see far right).

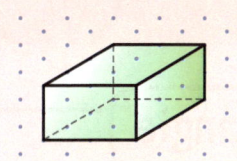

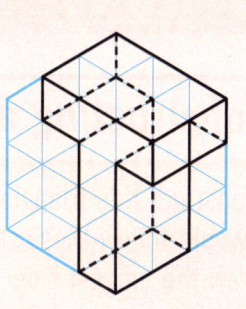

Nets of solids

The **net** of a 3D shape is the 2D (flat) shape that can be folded to make the 3D shape. For example:

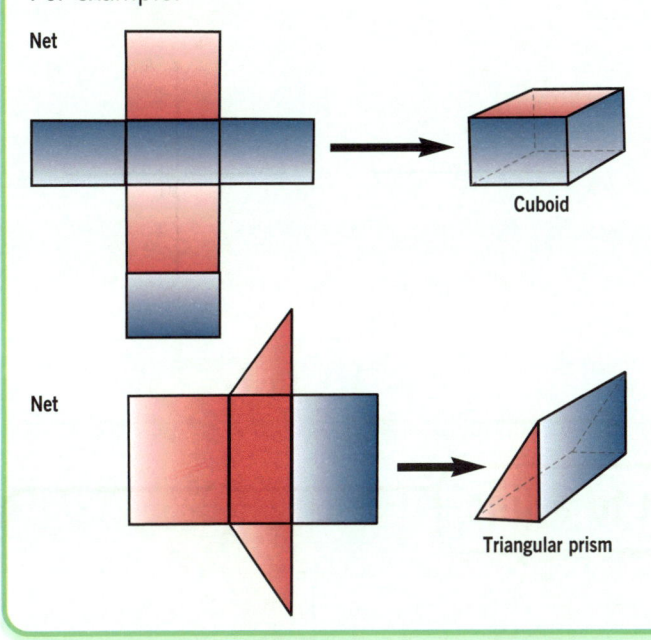

Net

Cuboid

Net

Triangular prism

When asked to draw an accurate net, you must measure carefully.

When making the shape, remember to put tabs on, to stick it together.

In manufacturing, the surface area of the net needs to be kept to a minimum in order to reduce costs.

Quick test

1. Draw an accurate net of this 3D shape.

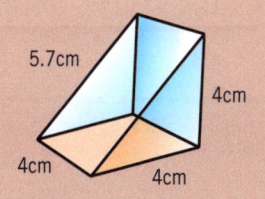

   5.7cm, 4cm, 4cm, 4cm

2. Draw a sketch of the plan, and elevations from A and B, of this solid.

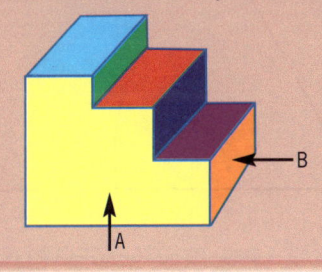

Geometry and measures

51

Constructions

Constructing a triangle

To construct this triangle using a compass and a ruler, you would follow these steps:
1. Draw the longest side.
2. With the compass point at A, draw an arc of radius 4cm.
3. With the compass point at B, draw an arc of radius 5cm.
4. Join A and B to the point where the two arcs meet at C.

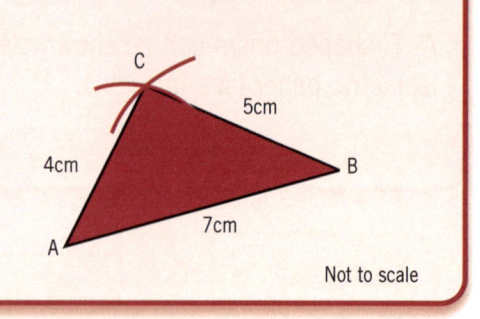

Not to scale

The perpendicular bisector of a line segment

To construct the **perpendicular bisector** of the **line segment** XY, you would follow these steps:
1. Draw two arcs with a compass, using X as the centre. The compass must be set at a radius greater than half the distance of XY.
2. Draw two more arcs with Y as the centre. (Keep the compass the same distance apart as before.)
3. Join the two points where the arcs cross.
4. AB is the **perpendicular bisector** of XY.
5. N is the **midpoint** of XY.

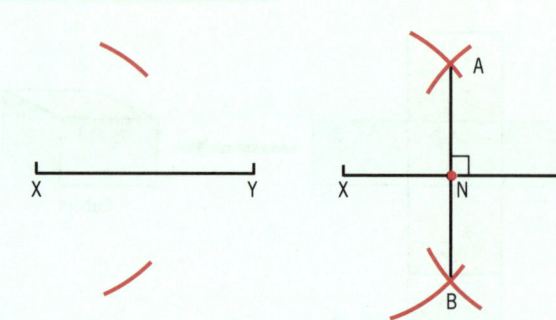

The perpendicular from a point to a line

To construct the perpendicular from point P to the line AB, you would follow these steps:
1. From P draw arcs to cut the line at A and B.
2. From A and B draw two arcs of the same radius to intersect at a point C below the line.
3. Join P to C; this line is perpendicular to AB.

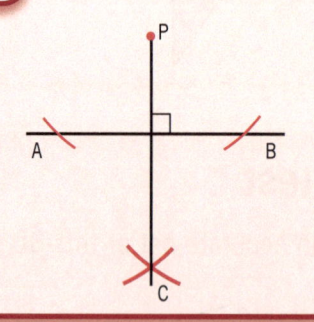

Constructing an angle of 60°

To construct an angle of 60° at point P on a line segment XY, you would follow these steps:
1. Put the compass point on P and draw an arc.
2. Label, with an N, the point where the arc intersects the line segment XY.
3. Keeping the radius the same, from point N draw an arc to intersect the first arc at point R.
4. Join PR. Angle RPN is 60°.

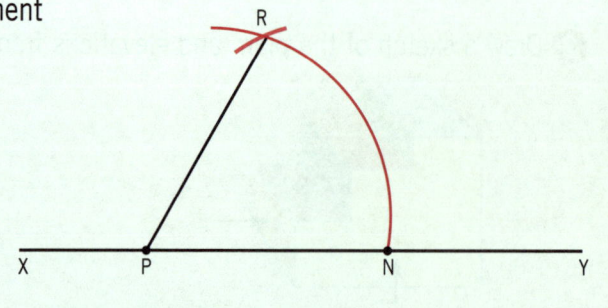

The perpendicular from a point on a straight line

To construct the perpendicular at the point N on a straight line, you would follow these steps:

1. With the compass set to a radius of several centimetres, and centred on N, draw arcs to cut the line at A and B.
2. Construct the **perpendicular bisector** of the line segment AB as on page 52.

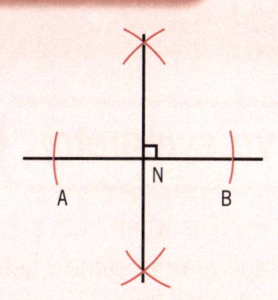

Bisecting an angle

To bisect an angle, follow these steps:
1. Draw two lines, XY and YZ, to meet at an angle.
2. Using a compass, place the point at Y and draw the two arcs on XY and YZ.
3. Place the compass point at the two arcs on XY and YZ, and draw arcs to cross at N.
4. Join Y and N. YN is the **bisector** of angle XYZ.

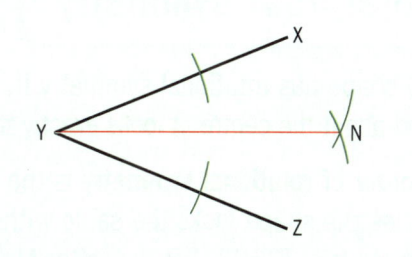

Construction of an inscribed regular polygon

Example
Construct a regular hexagon inside a circle of radius 2cm.

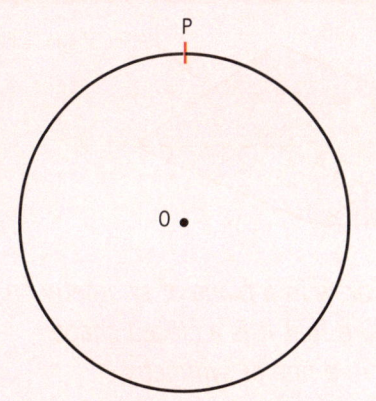

Draw a circle of radius 2cm and mark a point P on its circumference.

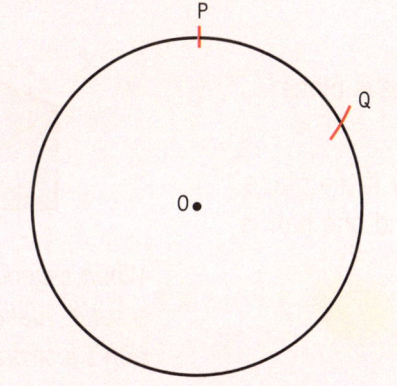

Keeping the compass set at 2cm, draw an arc, centre P, which cuts the circle at Q. Q is the centre of the next arc.

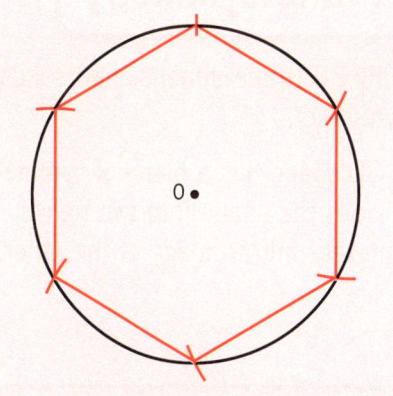

Repeat the process until six points are marked on the circumference. Join the points to make a hexagon.

Quick test

1. Bisect angle $x\ y\ z$.
2. Draw the perpendicular bisector of a 10cm line.

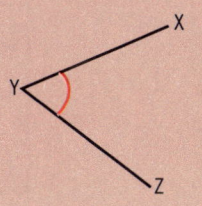

Diagram drawn to scale

Geometry and measures

53

Symmetry

Reflective symmetry

Reflective symmetry is when both sides of a symmetrical shape are the same when the mirror line is drawn across it. The mirror line is known as the **line** or **axis of symmetry**.

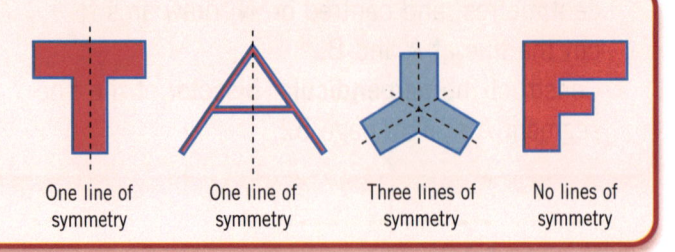

One line of symmetry | One line of symmetry | Three lines of symmetry | No lines of symmetry

Rotational symmetry

A 2D shape has rotational symmetry if, when turned about the centre, it looks exactly the same.

The **order of rotational symmetry** is the number of times the shape looks the same within one complete turn (360°). For the letter M, the shape has 1 such position. It is said to have rotational symmetry of order 1, or no rotational symmetry. The same is true for the letter T.

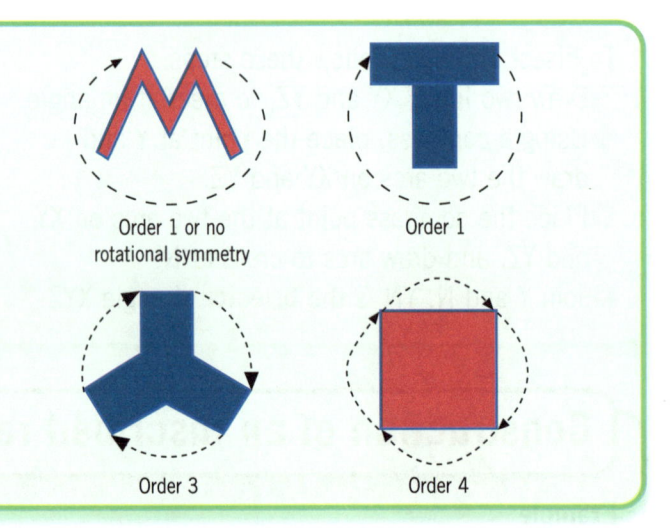

Order 1 or no rotational symmetry | Order 1 | Order 3 | Order 4

Plane symmetry

Only 3D (three-dimensional) solids have plane symmetry.

A 3D shape has a **plane of symmetry** if the plane divides the shape into two halves, and one half is an exact mirror image of the other.

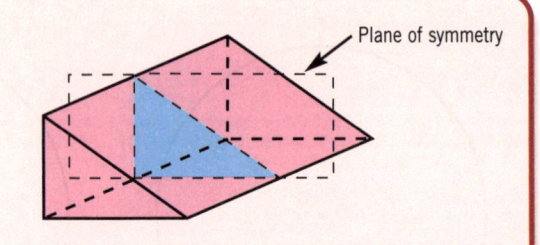

Plane of symmetry

When asked to draw in a plane of symmetry on a solid, make sure that it is a closed shape – don't just draw in a line of symmetry.

Quick test

1. What are the three types of symmetry?
2. The dashed lines on the diagram opposite are the lines of symmetry.
   Complete the shape so that it is symmetrical.
3. Draw a plane of symmetry on this solid.

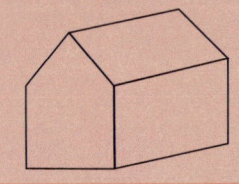

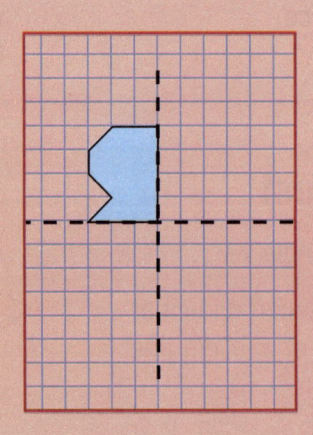

Geometry and measures

Loci & coordinates in 3D

Common loci

The **locus** of a point is the set of all the possible positions which that point can occupy, subject to some given condition or rule. The plural of locus is **loci**.

Remember these points:

The locus of the points that are equidistant from a fixed point P is a circle.	The locus of the points that are equidistant from two non-parallel lines is the line that bisects the angle formed by the two lines.
The locus of the points that are equidistant from two points, X and Y, is the perpendicular bisector of XY.	The locus of the points that are a constant distance from a line is a pair of parallel lines above and below the line.

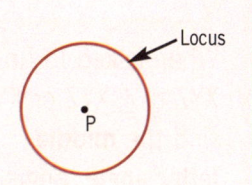

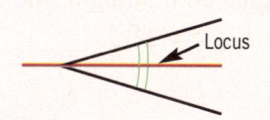

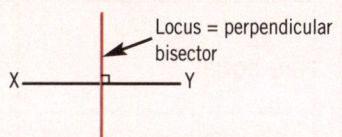

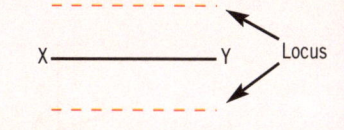

When answering loci questions, use the construction techniques previously shown: this will ensure your work is accurate.

Coordinates in 3D

Coordinates in 3D involve the extension of the normal x and y-axes into a third direction, known as the z-axis. All positions then have three coordinates (x, y, z).

For example, for the cuboid opposite the vertices would have the following (x, y, z) coordinates:

A (3, 0, 0)   E (0, 0, 1)
B (3, 2, 0)   F (3, 0, 1)
C (0, 2, 0)   G (3, 2, 1)
D (0, 0, 0)   H (0, 2, 1)

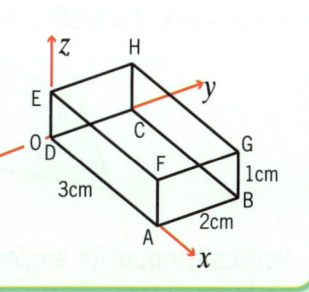

Quick test

1. A gold coin is buried in a rectangular field. The coin is 4m from T and equidistant from RU and RS. Copy the diagram to a scale of 1cm : 1m and mark with an X the position of the gold coin.

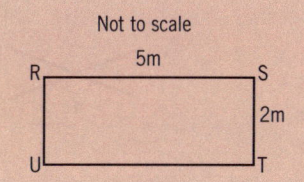

Angles

Types of angles

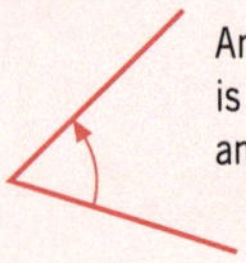

An **acute angle** is between 0° and 90°.

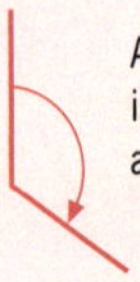

An **obtuse angle** is between 90° and 180°.

A **reflex angle** is between 180° and 360°.

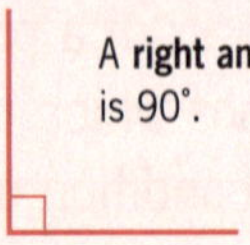

A **right angle** is 90°.

Angle facts

Angles on a straight line add up to **180°**.
$a + b + c = 180°$

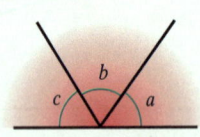

Angles at a point add up to **360°**.
$a + b + c + d = 360°$

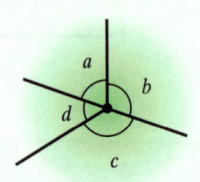

Angles in a triangle add up to **180°**.
$a + b + c = 180°$

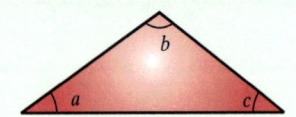

Angles in a quadrilateral add up to **360°**.
$a + b + c + d = 360°$

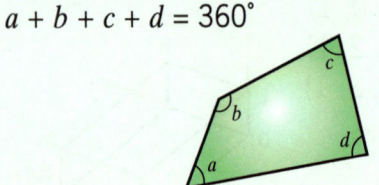

Vertically-opposite angles are **equal**.
$a = b$, $c = d$
$a + d = b + c = 180°$

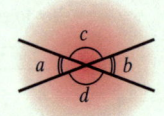

Reading angles

When asked to find XYZ or ∠XYZ or XŶZ, find the **middle letter angle**, angle Y.

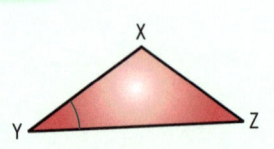

Angles in parallel lines

Alternate angles are **equal**.	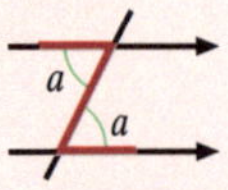
Corresponding angles are **equal**.	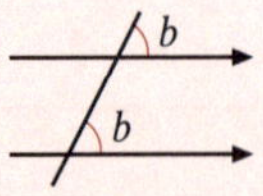
Supplementary angles add up to **180°**: $c + d = 180°$	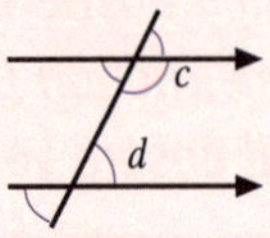

Examples
Find the angles labelled by letters.

a) $a = 180° - 50° - 70° = 60°$
$b = 180° - 60° = 120°$

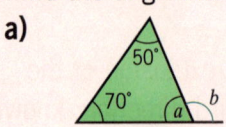

b) $a + 80° + 40° + 85° = 360°$
$a = 360° - 205°$
$a = 155°$

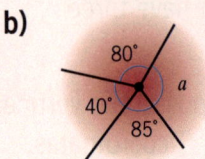

c) $a = 120°$ (angles on a straight line)
$b = 60°$ (vertically opposite to 60°)
$c = 60°$ (corresponding to b or alternate to 60°)
$d = 60°$ (vertically opposite to c)

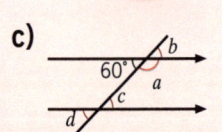

Measuring angles

A protractor is used to measure the size of an angle.

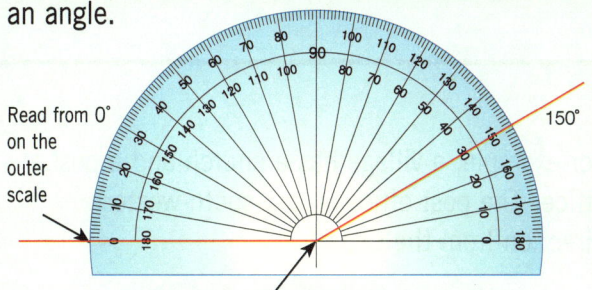

Read from 0° on the outer scale

Place the cross at the point of the angle you are measuring

For the above angle, measure on the outer scale since you must start from 0°.

> *Make sure you put the 0° line at the start position and read from the correct scale. When measuring angles, count the degree lines carefully and always double check.*

Tessellations

A **tessellation** is a pattern of 2D shapes that fit together without leaving any gaps.

For shapes to tessellate, the angles at each point must add up to 360°.

Regular pentagons will not tessellate. Each interior angle is 108°, and 3 × 108° = 324°. A gap of 360° – 324° = 36° is left.

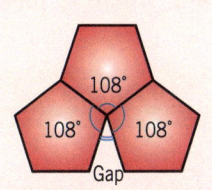

M.C. Escher is a famous artist who used tessellations in his art work. Tessellations are often seen in mosaics and floor tiles.

Angles in a polygon

There are two types of angle in any polygon: **interior** (inside) and **exterior** (outside).

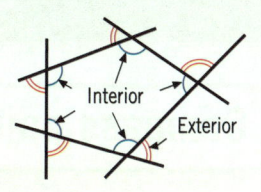

For a regular polygon with n sides…
- sum of exterior angles = 360°, so each exterior angle = $\dfrac{360°}{n}$
- interior angle + exterior angle = 180°
- sum of interior angles = $(n - 2) \times 180°$ or $(2n - 4) \times 90°$

Examples

a) A regular polygon has an interior angle of 150°. How many sides does it have?

Let n be the number of sides.
Exterior + interior = 180°
Exterior angle = 180° – 150° = 30°
Exterior angle = $\dfrac{360°}{n}$
So $n = \dfrac{360°}{\text{exterior angle}} = \dfrac{360°}{30°} = 12$

b) The diagram shows an irregular pentagon. Work out the size of x.

Sum of interior angles:
$(5 - 2) \times 180°$
$= 3 \times 180° = 540°$

Form an equation and solve it to find x:
$x + 2x + x + 30° + 3x + 3x + 10° = 540°$
$10x + 40° = 540°$
$10x = 540° - 40°$
$10x = 500°$, so $x = 50°$

> *Make sure that you show full working out when carrying out an angle calculation. If you are asked to 'Explain', always refer to the angle properties, e.g. angles on a straight line add up to 180°.*

Quick test

1. Find the sizes of the angles labelled by letters.
   a) 30°, a
   b) b, c, d, 110°
   c) 50°, a, b, c, d

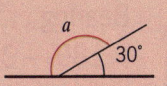

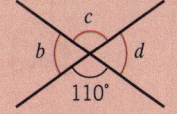

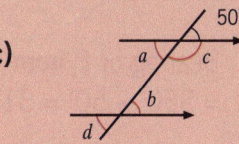

2. Find the size of an **a)** exterior and **b)** interior angle of a regular pentagon.

Geometry and measures

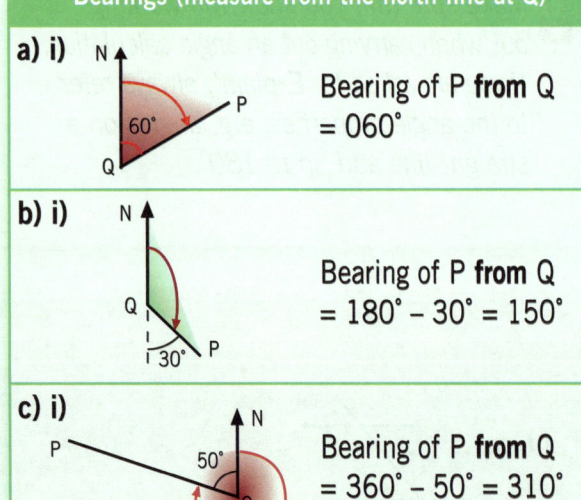

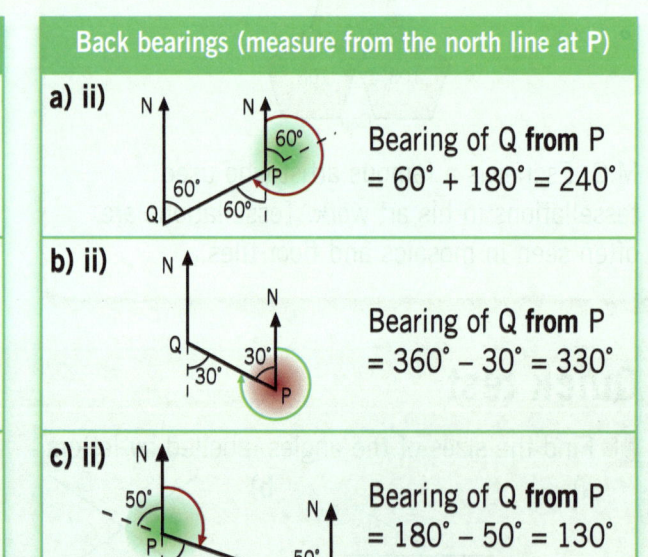

Bearings & scale drawings

Compass points

The diagram shows the points of the compass. Compass points tell us the direction of something.

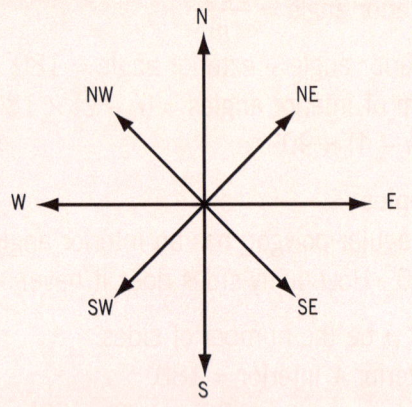

For example, a village has a church and a post office. The post office is in a south-westerly direction from the church.

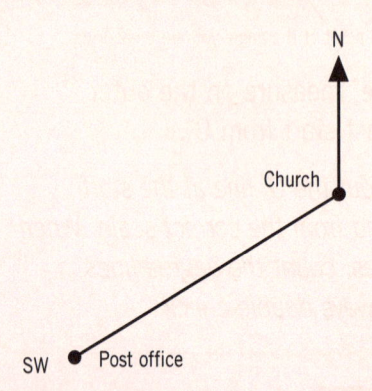

Bearings

A **bearing** is the direction travelled between two points, given as an angle in degrees:
- All bearings are measured clockwise from the north line.
- All bearings should be given as three figures, e.g. 225°, 043°, 006°.

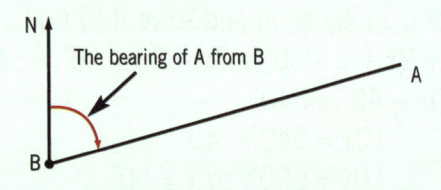

Back bearings
To find the **back bearing** (the bearing of Q from P in the examples below):
1. Draw in a north line at P.
2. The two north lines are parallel lines, so the angle properties of parallel lines can be used.

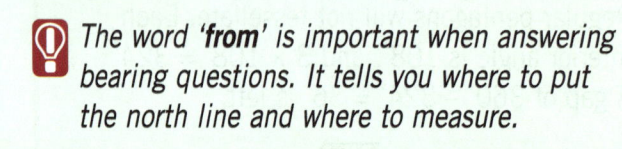

The word 'from' is important when answering bearing questions. It tells you where to put the north line and where to measure.

Bearings (measure from the north line at Q)	Back bearings (measure from the north line at P)
a) i) Bearing of P from Q = 060°	a) ii) Bearing of Q from P = 60° + 180° = 240°
b) i) Bearing of P from Q = 180° − 30° = 150°	b) ii) Bearing of Q from P = 360° − 30° = 330°
c) i) Bearing of P from Q = 360° − 50° = 310°	c) ii) Bearing of Q from P = 180° − 50° = 130°

Scale drawings and bearings

Scale drawings are useful for finding lengths and angles. They are often used by architects and surveyors.

Example
A ship sails from a harbour for 15km on a bearing of 040°, then continues due east for 20km. Make a scale drawing of this journey using a scale of 1cm to 5km. How far will the ship have to sail to get back to the harbour by the shortest route? What will the bearing be?

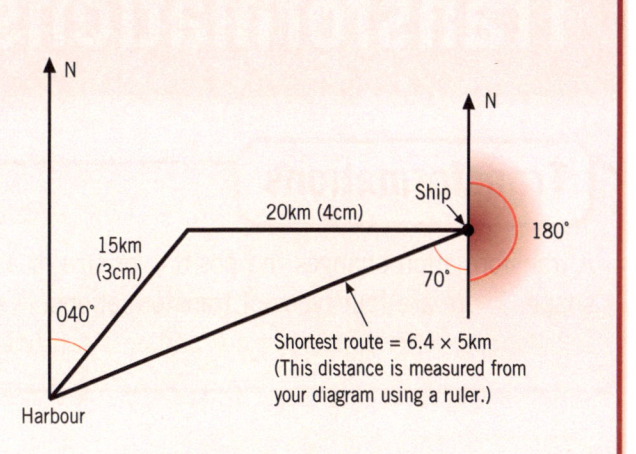

Shortest route = 6.4 × 5km = 32km
Bearing = 70° + 180° = 250°

Shortest route = 6.4 × 5km
(This distance is measured from your diagram using a ruler.)

Note – this diagram is not drawn to scale but is used to show you what your diagram should look like.

Scales and maps

Scales are often used on maps. They are usually written as a ratio. You will do map work in geography. Orienteers use maps and compasses to find directions.

A scale of 1 : 25 000 means that 1cm on the scale drawing represents a real length of 25 000cm.

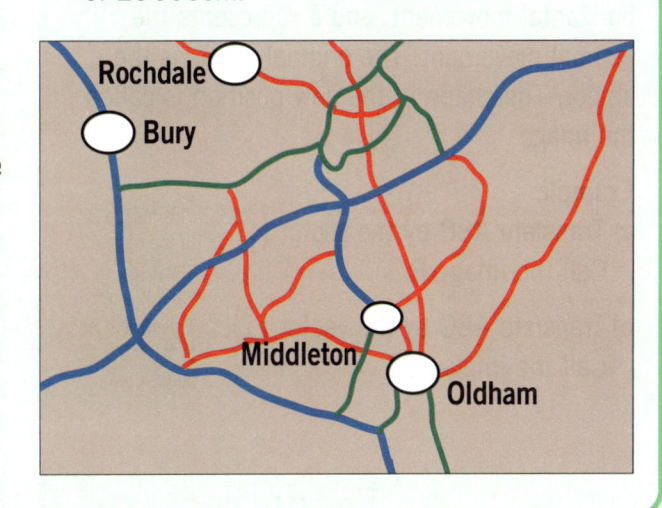

Example
The scale on a road map is 1 : 25 000. Bury and Oldham are 60cm apart on the map. Work out the real distance, in km, between Bury and Oldham.

① Scale 1 : 25 000, distance on map is 60cm.
∴ Real distance = 60 × 25 000 = 1 500 000cm
② Divide by 100 to change cm to m:
1 500 000 ÷ 100 = 15 000m
③ Divide by 1000 to change m to km:
15 000 ÷ 1000 = 15km

Quick test

① What are the bearings of A from B in the following diagrams?
a) 72°
b) 55°
c) 35°

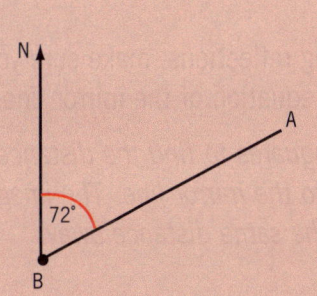

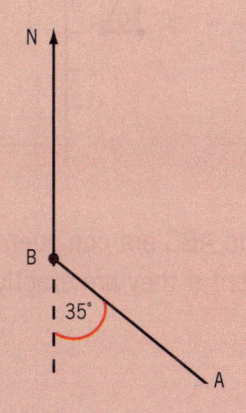

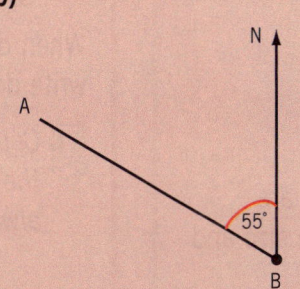

② For each of the questions above, work out the bearing of B from A.

③ The scale on a road map is 1 : 50 000. If two towns are 14cm apart on the map, work out the real distance between them in kilometres.

Geometry and measures

Transformations 1

Transformations

A transformation changes the position or size of a shape. There are four types of transformations: **translations**, **reflections**, **rotations** and **enlargements**.

Transformations are often used in the creation of repeating wallpaper patterns.

Translations

A **translation** moves a figure from one place to another. The size, shape and orientation of the figure are not changed. **Vectors** are used to describe the distance and direction of a translation.

A vector is written $\binom{a}{b}$, where a represents the **horizontal** movement, and b represents the **vertical** movement. The original shape is the object. The shape in the new position is called the **image**.

Example
a) Translate ABC by the vector $\binom{2}{1}$. Call the image P.

This means 2 to the right and 1 up.

b) Translate ABC by the vector $\binom{-3}{-2}$. Call the image Q.

This means 3 to the left and 2 down.

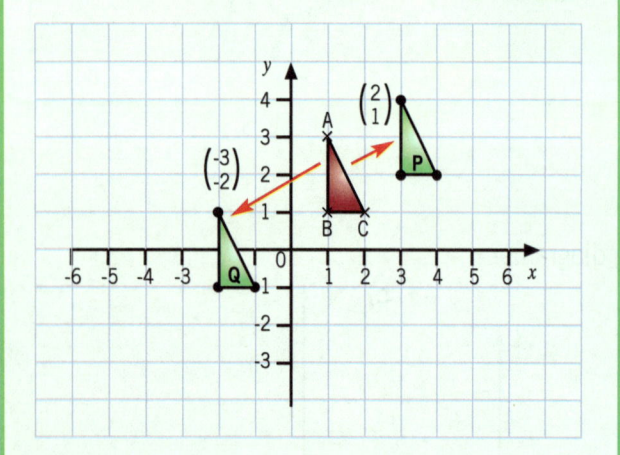

P, Q and ABC are **congruent** – shapes are congruent if they are exactly the same size and shape.

Reflections

A **reflection** creates an image of an object on the other side of the **mirror line**. The size and shape of the figure are not changed.

Example
Reflect triangle ABC in...
a) the x-axis, and call the image D
b) the line $y = -x$, and call the image E
c) the line $x = 5$, and call the image F.

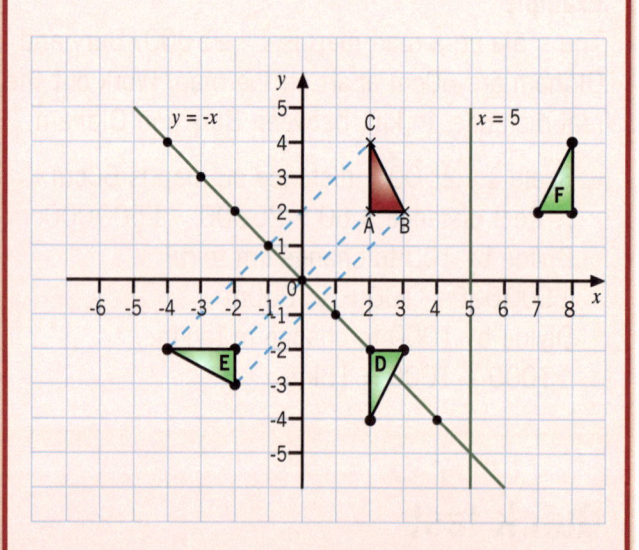

Triangles D, E and F are congruent to triangle ABC.

When describing reflections, make sure you write down the equation of the mirror line.

⚠ *Count the squares to find the distance from the object to the mirror line. The image will always be the same distance away.*

Rotations

A **rotation** turns a figure through an angle about some fixed point. This fixed point is called the **centre of rotation**. The size and shape of the figure are not changed.

Example
Rotate triangle ABC...
a) 90° clockwise about (0, 0), and call it R
b) 180° about (0, 0), and call it S
c) 90° anticlockwise about (-1, 1), and call it T.

Triangles R, S and T are all congruent to triangle ABC.

When describing a rotation, give...
- the centre of rotation
- the direction of the turn (clockwise or anticlockwise)
- the angle of the turn.

If you do not give all three pieces of information, you will lose marks for not describing the rotation fully.

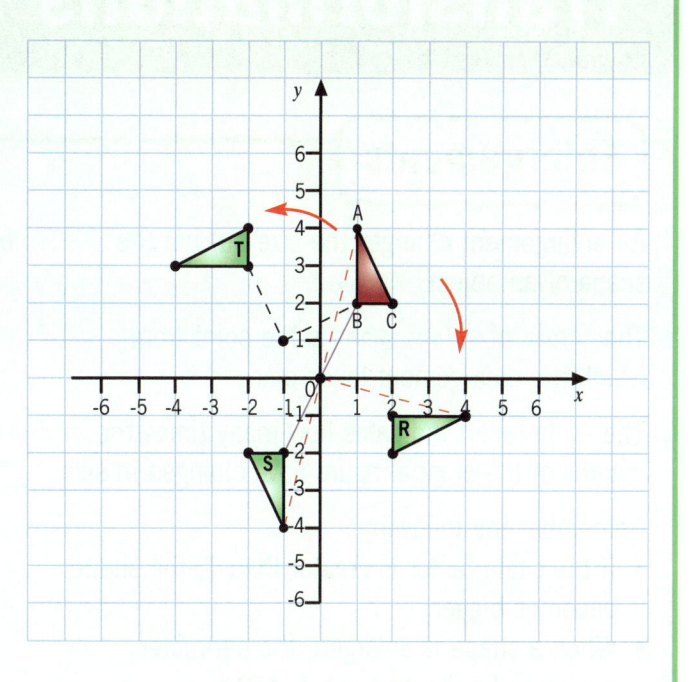

Quick test

1. On the diagram opposite...
   a) translate triangle ABC by the vector $\binom{-3}{1}$, and call it P
   b) reflect ABC in the line $y = x$, and call it Q
   c) reflect ABC in the line $y = -1$, and call it R
   d) rotate ABC 180° about (0, 0), and call it S.

2. What does the vector $\binom{-2}{3}$ mean?

3. A point R (3, -6) moves to the point T (-2, 5). What is the vector translation that moves point R to point T?

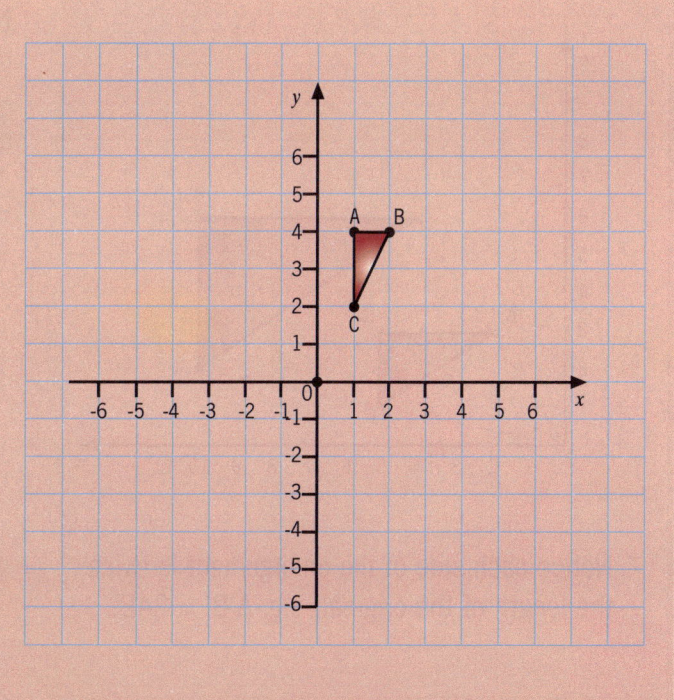

Transformations 2

Enlargements

An **enlargement** changes the size but not the shape of an object.

The **centre of enlargement** is the point from which the enlargement takes place.

The **scale factor** indicates how many times the lengths of the original figure have changed in size.

Remember the following:
- If the scale factor is **greater than 1**, the shape becomes **bigger**.
- When a shape is enlarged, the perimeter increases by the same scale factor.
- When one shape is an enlargement of another, they are **similar**, i.e. they are the same shape with the same angles, but they are different sizes.

Examples

a) Enlarge triangle ABC by a scale factor of 2, centre = (0, 0). Call it A'B'C'.

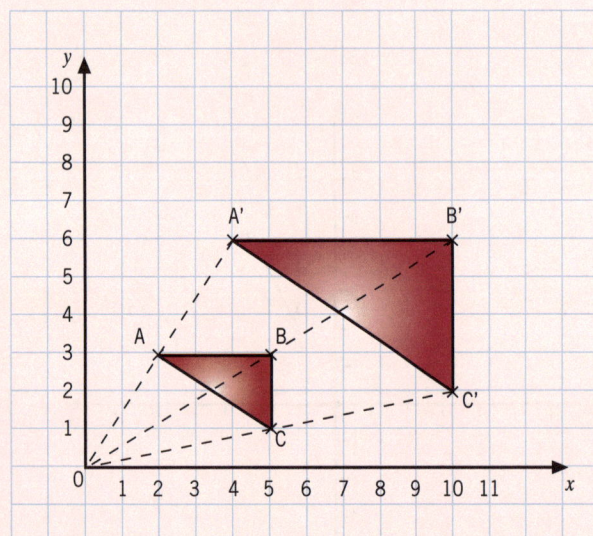

b) Describe fully the transformation that maps ABCDEF onto A'B'C'D'E'F'.

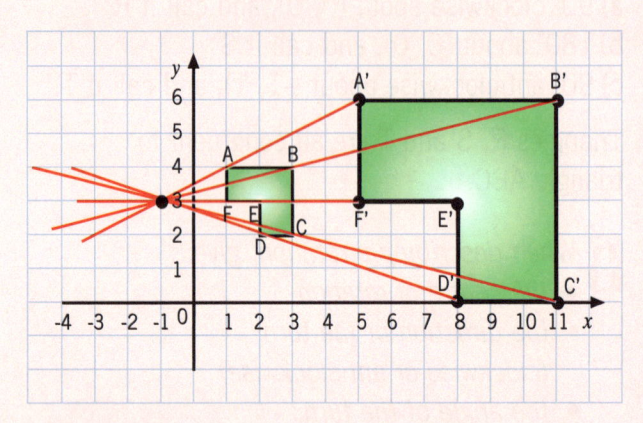

- To find the centre of enlargement, join A to A', and continue the line. Join B to B', and continue the line. Do the same for the others.
- Where all the lines meet is the centre of enlargement: (-1, 3).
- The transformation is an enlargement with scale factor 3 and centre of enlargement at (-1, 3).

Notice that A'B'C'D'E'F' is three times the size of ABCDEF.

When asked to describe an enlargement, you must include the scale factor and the position of the centre of enlargement.

Notice each side of the enlargement is twice the length of the original, e.g. A'B' = 2AB.

Combining transformations

Combined transformations are a series of two or more transformations.

Example

a) Reflect triangle ABC in the *x*-axis and call the image $A_1B_1C_1$.

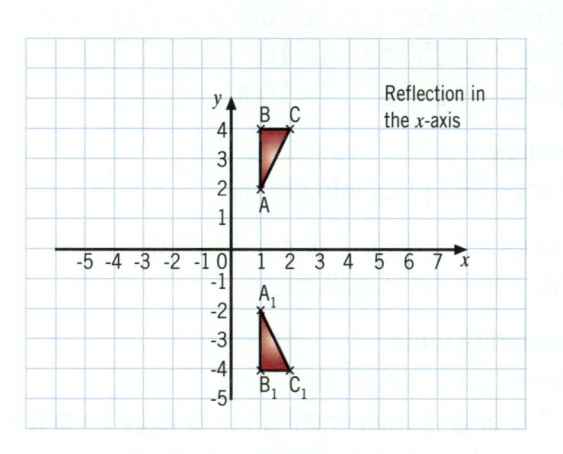

b) Reflect triangle $A_1B_1C_1$ in the *y*-axis and call the image $A_2B_2C_2$.

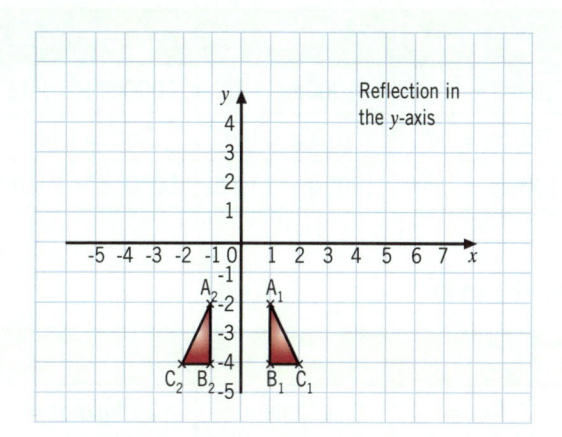

Reflection in the *x*-axis

+

Reflection in the *y*-axis

The single transformation that maps ABC directly onto $A_2B_2C_2$ is a rotation of 180° about centre (0, 0).

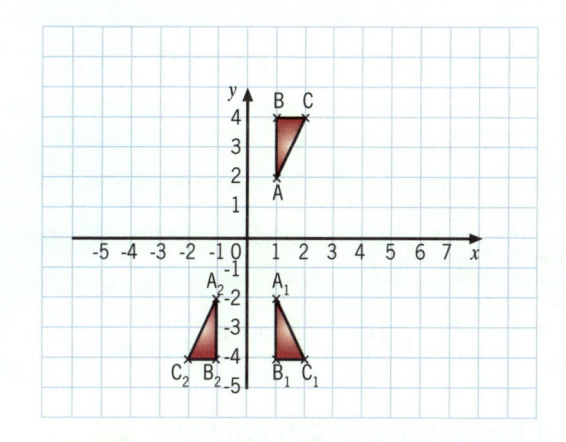

=

Quick test

1. Draw an enlargement of shape P with a scale factor of 2. Call it P1.
2. **a)** Rotate the shaded shape through a 90° clockwise rotation about (0, 0). Call this shape B.
   b) Reflect shape B in the *x*-axis. Call this shape C.

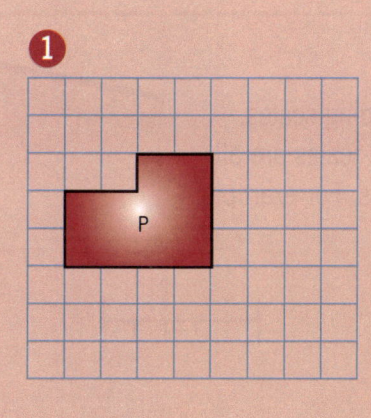

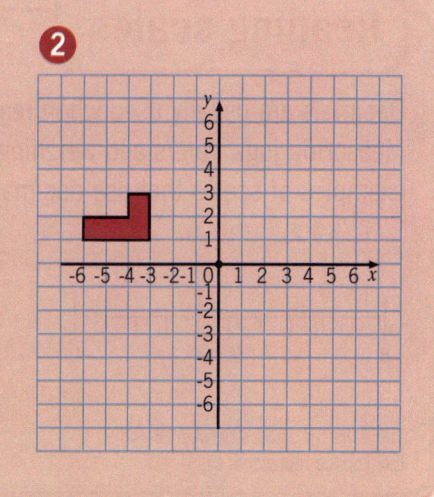

Measures & measurement 1

Metric and imperial units

Metric units

Length	Weight	Capacity
10mm = 1cm	1000mg = 1g	1000ml = 1 litre
100cm = 1m	1000g = 1kg	100cl = 1 litre
1000m = 1km	1000kg = 1 tonne	1000cm^3 = 1 litre

Imperial units

Length	Weight	Capacity
1 foot = 12 inches	1 stone = 14 pounds (lb)	20 fluid oz = 1 pint
1 yard = 3 feet	1 pound = 16 ounces (oz)	8 pints = 1 gallon

Here are some metric conversions:
500cm = 5m (÷ 100)
25cm = 250mm (× 10)
3500g = 3.5kg (÷ 1000)

Example
Change 166 pints into gallons.

8 pints = 1 gallon, 1 pint = $\frac{1}{8}$ gallon
166 ÷ 8 = 20.75 gallons

Converting units
Remember these points:
- If changing from **small** units **to large** units (for example, grams to kilograms), **divide**.
- If changing from **large** units **to small** units (for example, kilometres to metres), **multiply**.

Comparisons between metric and imperial units

Length	Weight	Capacity
2.5cm ≈ 1 inch	25g ≈ 1 ounce	1 litre ≈ 1$\frac{3}{4}$ pints *
30cm ≈ 1 foot *		
1m ≈ 39 inches	1kg ≈ 2.2 pounds *	4.5 litres ≈ 1 gallon *
8km ≈ 5 miles *		

All of the comparisons between metric and imperial units are only approximate.

Example
Change 25km into miles.

8km ≈ 5 miles
1km ≈ $\frac{5}{8}$ mile = 0.625 miles
25km ≈ 25 × 0.625
= 15.625 miles

*There is a lot of learning to do in this section. Try to learn all the metric and imperial conversions marked *.*

Reading scales

Decimals are usually used when reading scales. Measuring jugs, rulers and weighing scales are all examples of scales that use decimals. For example:

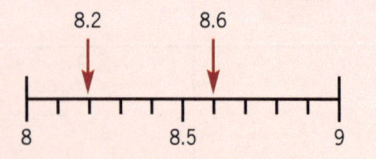

There are ten spaces between 8 and 9. Each space represents 0.1

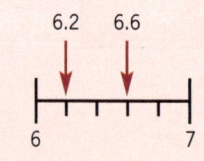

There are five spaces between 6 and 7. Each space represents 0.2

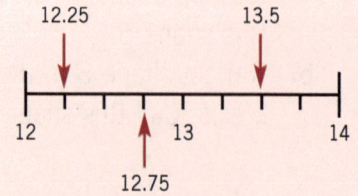

There are eight spaces between 12 and 14. Each space represents 0.25

Estimating measures

Estimates using metric and imperial units need to be made all the time.

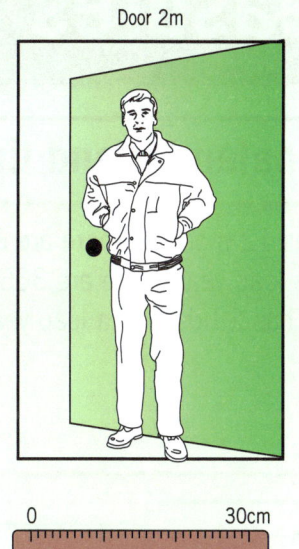

Estimating lengths
Length and distance can be measured in the following units:
- **Metric:** kilometres (km), metres (m), centimetres (cm) and millimetres (mm).
- **Imperial:** inches, feet, yards and miles.

Some common estimates include:
- A door is about 2 metres high or about $6\frac{1}{2}$ feet.
- A 30cm ruler is about 1 foot long.

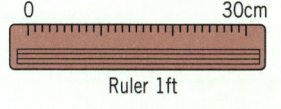

Estimating capacities
Capacity is a measure of how much a container can hold. It can be measured in the following units:
- **Metric:** millilitres (ml), centilitres (cl) and litres (l).
- **Imperial:** pints and gallons.

Some common estimates include:
- A 1 pint milk carton holds about 570ml.
- A petrol can holds 1 gallon or 4.5 litres.
- A can of pop holds about 300ml or $\frac{1}{2}$ pint.

Estimating weights (masses)
Weight can be measured in the following units:
- **Metric:** milligrams (mg), grams (g), kilograms (kg) and tonnes (t).
- **Imperial:** ounces (oz), pounds (lb), stones (st) and tons.

Some common estimates include:
- A 1kg bag of sugar weighs about 2.2lb.
- A 250g packet of butter weighs about $\frac{1}{2}$lb.

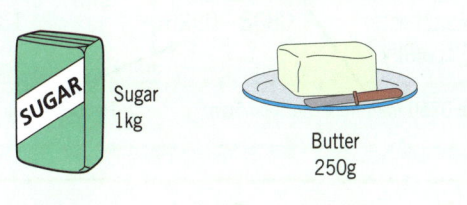

Quick test

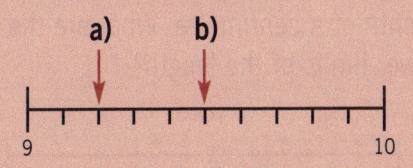

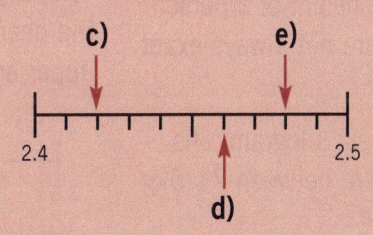

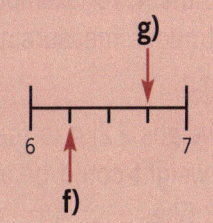

1. Change 3500g into kilograms.
2. Change 3kg into pounds.
3. Change 8 litres into pints.
4. What do the pointers on the scales represent?
5. What would be the approximate weight of a newborn baby?
   A   50g   B   5g   C   3kg   D   30kg

Measures & measurement 2

The calendar and time facts

A year has 12 months. There are 365 days in a year. In a leap year, there are 366 days because February has 29 days in a leap year.

Here are some other important time facts. There are...
- 60 seconds in 1 minute
- 60 minutes in 1 hour
- 24 hours in 1 day
- 7 days in 1 week
- 52 weeks in 1 year.

Timetables

Timetables often use the 24-hour clock. Timetables should be read carefully.

Example
This train timetable shows the train times from London to Manchester.

There will be a train from London every 60 minutes (or 1 hour), i.e. 0750, 0850, etc.

London Euston	0602	0650		1100	1300
Watford Junction	0632	0720	Every 60 minutes until	1130	1330
Stoke-on-Trent	0750	0838		–	1445
Manchester Piccadilly	0838	0926		1315	1540

The 0750 train from Stoke-on-Trent.

The 0650 train from London arrives in Manchester at 0926.

The 1100 train from London does not stop at Stoke-on-Trent.

The 0632 train from Watford Junction takes 2 hours 6 minutes to travel to Manchester.

If a timetable is written in 24-hour clock times, make sure your answers are in 24-hour clock time.

Reading a timetable is a very important technique to learn. You will be expected to be able to read and interpret a selection of different timetables and charts.

Accuracy of measurement

There are two types of measurements – discrete measurements and continuous measurements:
- **Discrete measurements** are quantities that can be counted; for example, the number of baked bean tins on a shelf.
- **Continuous measurements** are measurements that have been taken with a measuring instrument; for example, the height of a person. Continuous measurements are **not always exact**.

Example
Nigel weighs 72kg to the nearest kilogram. His actual weight could be anywhere between 71.5kg and 72.5kg.

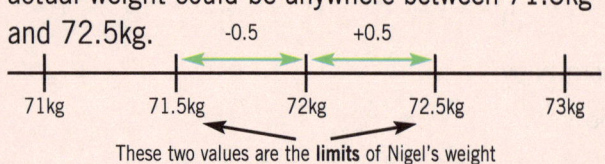

These two values are the **limits** of Nigel's weight.

If W represents Nigel's weight, then
$$71.5 \leq W < 72.5$$

This is the **lower limit** of Nigel's weight (sometimes known as the **lower bound**). Anything below 71.5 would be recorded as 71kg.

This is the **upper limit** (**upper bound**) of Nigel's weight. Anything from 72.5 upwards would be recorded as 73kg.

Example
The length of a seedling is measured as 3.7cm to the nearest tenth of a centimetre. What are the upper and lower limits of the length?

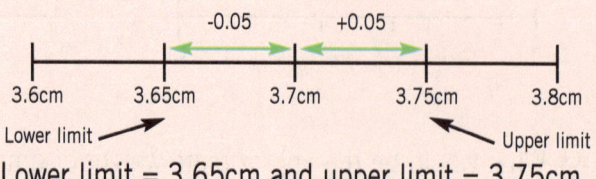

Lower limit = 3.65cm and upper limit = 3.75cm.

$$3.65 \leq L < 3.75$$

Compound measures

Speed can be measured in kilometres per hour (km/h), miles per hour (mph) and metres per second (m/s). These are all **compound measures** because they involve a combination of basic measures.

Speed

$$\text{Average speed} = \frac{\text{total distance travelled}}{\text{total time taken}}$$
$$s = \frac{d}{t}$$

Always check the units before starting a question. Change them if necessary.

Example
A car travels 50 miles in 1 hour 20 minutes. Find the speed in miles per hour.

Change the time units first:
20 minutes = $\frac{20}{60}$ of 1 hour
$s = \frac{d}{t} = \frac{50}{1\frac{20}{60}} = 37.5$ mph

From the speed formula, two other useful formulae can be found:

Time = $\frac{\text{distance}}{\text{speed}}$
$t = \frac{d}{s}$

Distance = speed × time
$d = st$

💡 *Just remember the letters: it's easier. Use the formula triangle to help you.*

$\triangle$ $\frac{d}{s \times t}$

Example
A car travels a distance of 240 miles at an average speed of 65mph.
How long does the journey take?

Time = $\frac{\text{distance}}{\text{speed}}$ so $t = \frac{240}{65} = 3.692$ hours

3.692 hours must be changed to hours and minutes.
- Subtract the hours. So 3.692 − 3 = 0.692
- Multiply the decimal part by 60 minutes.
0.692 × 60 = 42 minutes (nearest minute)
So time = 3 hours 42 minutes.

Density

Density = $\frac{\text{mass}}{\text{volume}}$ $D = \frac{M}{V}$

Volume = $\frac{\text{mass}}{\text{density}}$ $V = \frac{M}{D}$

Mass = density × volume $M = DV$

Example
Find the density of an object with mass 400g and volume 25cm³.

Density = $\frac{M}{V} = \frac{400}{25} = 16$ g/cm³

Since the mass is in grams and volume is in cm³, density is in g/cm³.

Quick test

1. What are the upper and lower limits for a time of 9.2 seconds, rounded to the nearest tenth of a second?
2. Write down the upper and lower limits for a weight of 58kg, rounded to the nearest kilogram.
3. Amy walks 6 miles in 2 hours 40 minutes. Find her average speed.
4. Find the time taken for a car to travel 600 miles at an average speed of 70mph.
5. Find the density of an object with mass 20g and volume 9cm³.

Pythagoras' theorem

The theorem

Pythagoras' theorem states: in any right-angled triangle, the square on the **hypotenuse** is equal to the sum of the squares on the other two sides.

The hypotenuse is the longest side of a right-angled triangle. It is always opposite the right angle.

Using the letters in the diagram, the theorem is written as $c^2 = a^2 + b^2$

This can be rearranged to give:
$a^2 = c^2 - b^2$ or $b^2 = c^2 - a^2$
These forms are useful when calculating the length of one of the shorter sides.

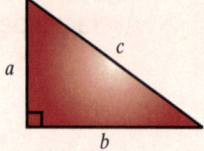

Finding the length of the hypotenuse

Remember: Pythagoras' theorem can only be used for right-angled triangles.

To find the length of the hypotenuse (longest side), follow these simple steps:
1. Square the two lengths of the two shorter sides that you are given.
2. To find the hypotenuse, **add** these two squared numbers.
3. After adding the two squared lengths, take the **square root** ($\sqrt{\ }$) of the sum.

Example
Find the length of AB, giving your answer to 1 decimal place.

Using Pythagoras' theorem:
$(AB)^2 = (AC)^2 + (BC)^2$
$= 12^2 + 14.5^2$
$= 354.25$
AB $= \sqrt{354.25}$ Take the square root to find AB.
$= 18.8$m   Round to 1 d.p.

Finding the length of a shorter side

To find the length of a shorter side, follow these steps:
1. Square the two lengths of the two sides you are given.
2. To find the shorter length, subtract the smaller value from the larger value.
3. Take the square root ($\sqrt{\ }$) of your answer.

💡 *Pythagoras' theorem allows us to calculate the length of one of the sides of a right-angled triangle when the other two sides are known. If you are not told to what degree of accuracy to round your answer, be guided by significant figures given in the question.*

Example
Find the length of FG, giving your answer to 1 decimal place.

Using Pythagoras' theorem:
$(EF)^2 = (EG)^2 + (FG)^2$
$(FG)^2 = (EF)^2 - (EG)^2$
$= 9^2 - 8^2$
$= 81 - 64$
$= 17$
FG $= \sqrt{17}$
$= 4.1$cm

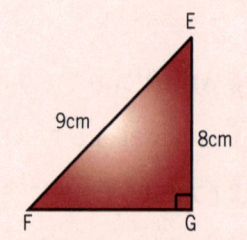

Geometry and measures

Calculating the length of a line AB, given two sets of coordinates

By drawing in a triangle between the two points A (1, 2) and B (7, 6) we can find the length of AB by using Pythagoras' theorem.

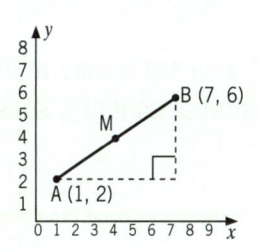

Horizontal distance = 6   (7 − 1)
Vertical distance = 4   (6 − 2)

Length of $(AB)^2 = 6^2 + 4^2$
$= 36 + 16$
$= 52$
$AB = \sqrt{52}$
Length of AB = 7.21 (2 d.p.)

The midpoint of AB, M, has coordinates
$\left(\dfrac{1+7}{2}, \dfrac{2+6}{2}\right) = (4, 4)$

Solving problems

Examples

a) Calculate the height of this isosceles triangle.

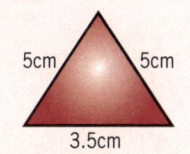

5cm, 5cm, 3.5cm

① Split the triangle down the middle to make it right-angled.

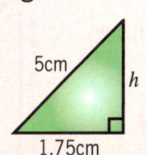

5cm, h, 1.75cm

② Using Pythagoras' theorem:
$5^2 = h^2 + 1.75^2$
$h^2 = 5^2 - 1.75^2$
$h^2 = 21.9375$
$h = \sqrt{21.9375} = 4.68$cm (2 d.p.)

b) A ladder of length 13m rests against a wall. The height that the ladder reaches up the wall is 12m. How far away from the wall is the foot of the ladder?

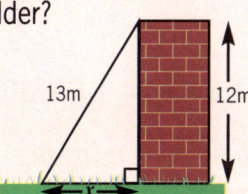

13m, 12m, x

$13^2 = x^2 + 12^2$
$x^2 = 13^2 - 12^2$
$x^2 = 169 - 144$
$x^2 = 25$
$x = \sqrt{25} = 5$m

The foot of the ladder is 5m away from the wall.

Quick test

① Calculate the lengths of the sides marked with the letter x. Give your answers to 1 decimal place.

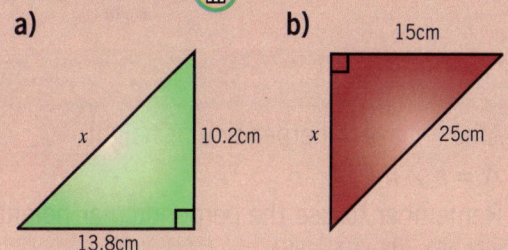

a) x, 10.2cm, 13.8cm

b) 15cm, x, 25cm

② Work out the length of the diagonal of this rectangle.

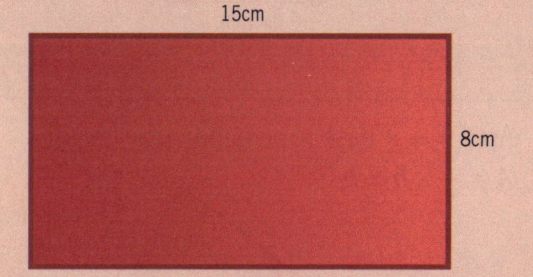

15cm, 8cm

③ A ship sets off from Port A and travels 50km due north, then 80km due east to reach Port B. How far is Port A from Port B by the shortest route? Give your answer to 1 d.p.

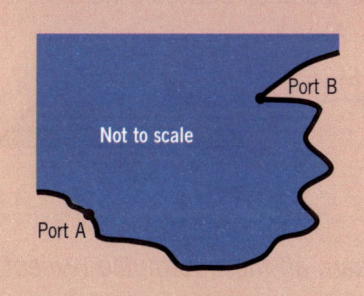

Not to scale

Geometry and measures

69

Area of 2D shapes

Perimeter and area of 2D shapes

The **perimeter** is the distance around the outside edge of a shape.

The **area** is the amount of space a 2D shape covers.

Common units of area are square millimetres (mm^2), square centimetres (cm^2), square metres (m^2), etc.

Areas of quadrilaterals and triangles

Area of a rectangle

Area = length × width
$A = l \times w$

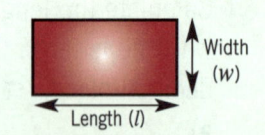

Area of a parallelogram

Area = base × perpendicular height
$A = b \times h$
Remember to use the perpendicular height, not the slant height.

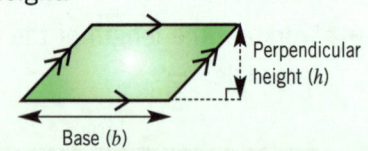

Area of a triangle

Area = ½ × base × perpendicular height
$A = \frac{1}{2} \times b \times h$

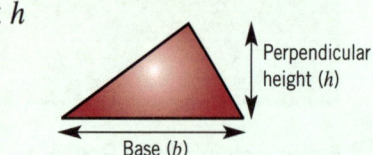

Area of a trapezium

Area = ½ × (sum of parallel sides) × perpendicular distance between them
$A = \frac{1}{2} \times (a + b) \times h$
or $A = \frac{(a+b)h}{2}$

❗ *You must learn all these formulae (except for the trapezium).*

Examples

a) Find the areas of the following shapes.

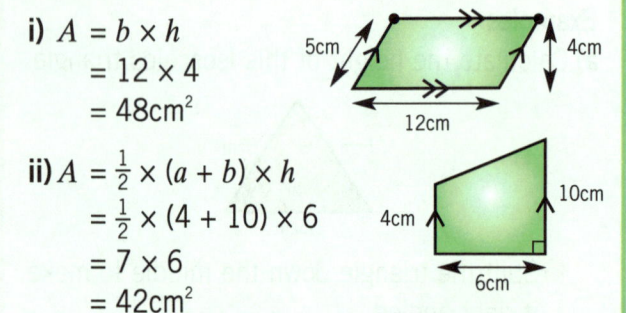

i) $A = b \times h$
   $= 12 \times 4$
   $= 48 cm^2$

ii) $A = \frac{1}{2} \times (a + b) \times h$
   $= \frac{1}{2} \times (4 + 10) \times 6$
   $= 7 \times 6$
   $= 42 cm^2$

b) The diagram opposite shows the plan of a floor.

i) Calculate the area of the floor.

   Split the shape into two parts:
   Rectangle: $A = l \times w$
   $= 6 \times 4$
   $= 24 m^2$

   Triangle: $A = \frac{1}{2} \times b \times h$
   $= \frac{1}{2} \times 4 \times 3$
   $= 6 m^2$

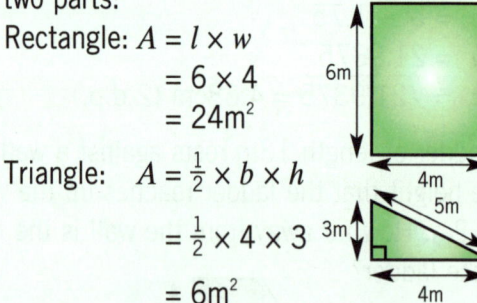

Total area of floor = 24 + 6 = $30 m^2$

❗ *This could also be done using the trapezium formula.*

ii) Skirting board is sold in 3m lengths. How many lengths are needed to go around the four walls of the room?

Perimeter of room = 3 + 6 + 4 + 6 + 5 = 24m
Skirting board = 3m lengths. 24 ÷ 3 = 8,
so eight lengths of skirting board are needed.

Areas of enlargement and changing area units

If a shape is enlarged by a **scale factor** n, then the **area** is n^2 times bigger. For example, if $n = 2$, the lengths are twice as big and the area is four times as big ($n^2 = 4$).

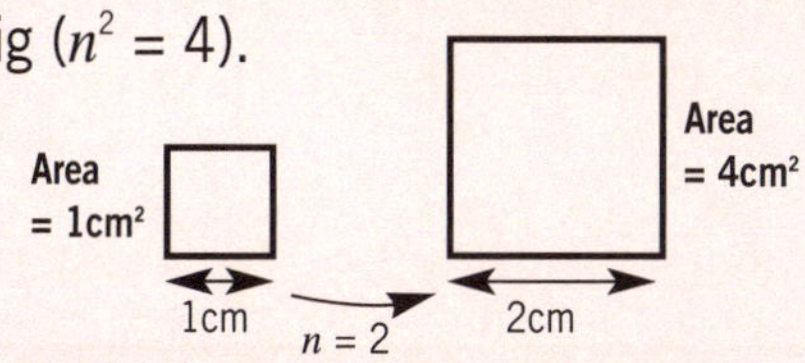

Area = 1cm²   Area = 4cm²
1cm $n = 2$ 2cm

💡 *It is always better to change units before you start a question in order to avoid possible errors.*

This example will help you to change area units. The square has a length of 1m. This is the same as a length of 100cm.

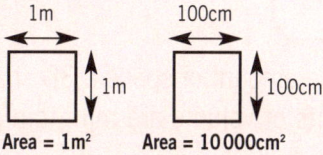

1m   100cm
1m   100cm
Area = 1m²   Area = 10 000cm²

Therefore 1m² = 10 000cm²
(not 100cm² as many people think!)

Circumference and area of a circle

Circumference = π × diameter or $C = \pi d$
Circumference = π × 2 × radius or $C = 2\pi r$

Area = π × (radius)²
$A = \pi r^2$

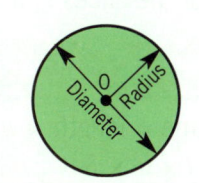

Examples

a) The diameter of a circular rose garden is 5m. Find the circumference and area of the garden. Use π = 3.142

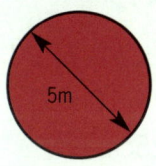

5m

$C = \pi \times d$
$= 3.142 \times 5$
$= 15.71$m

Substitute in the formula. Use π = 3.142 or the value of π on your calculator. EXP gives the value of π on some calculators.

When finding the area, work out the radius first.
$d = 2 \times r$, so $r = d \div 2$. Here $r = 2.5$m
$A = \pi \times r^2$
$= 3.142 \times 2.5^2$
$= 19.6375$
$= 19.6$m² (3 s.f.)

Remember 2.5² means 2.5 × 2.5

This answer could be left as 6.25π, i.e. in terms of π.

b) Work out the area of the semicircle. Leave your answer in terms of π.

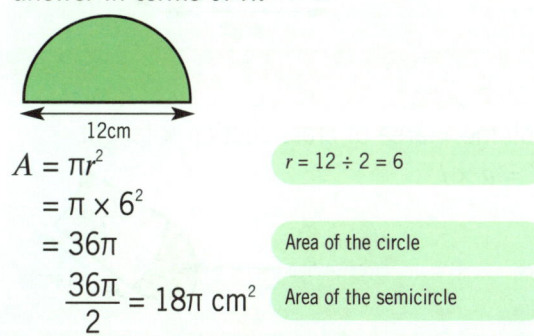

12cm

$A = \pi r^2$
$= \pi \times 6^2$
$= 36\pi$
$\dfrac{36\pi}{2} = 18\pi$ cm²

$r = 12 \div 2 = 6$

Area of the circle

Area of the semicircle

c) The circumference of a circle is 200cm. Work out the diameter of the circle. Use the π key.
$C = \pi \times d$
$200 = \pi d$
$\dfrac{200}{\pi} = d$
$d = 63.7$cm (3 s.f.)

Quick test

1 Work out the areas of these shapes, giving your answers to 3 significant figures where appropriate.

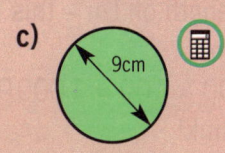

a) 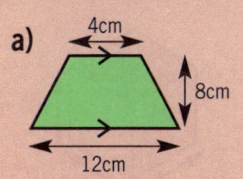 4cm, 8cm, 12cm
b) 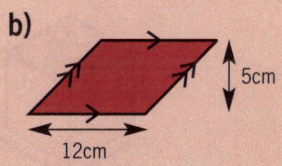 5cm, 12cm
c) 9cm
d) 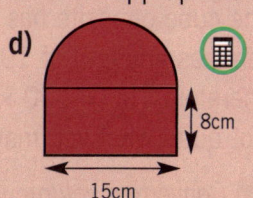 8cm, 15cm

2 Work out the area of the shaded region. Give your answer to 1 d.p.

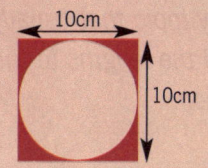

10cm, 10cm

Geometry and measures

71

Volume of 3D shapes

Volume

Volume is the amount of space a 3D shape occupies.
Common units of volume are mm³, cm³, m³, etc.

Volume of prisms

A **prism** is any solid that can be cut across its length into slices that are all the same shape. The shape of the slices is the uniform cross-section.

Volume of a cuboid

Volume = length × width × height
$V = l \times w \times h$

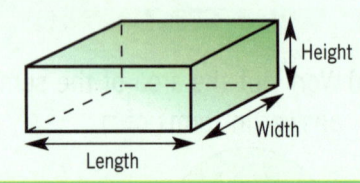

Volume of a prism

Volume = area of cross-section × length
$V = a \times l$

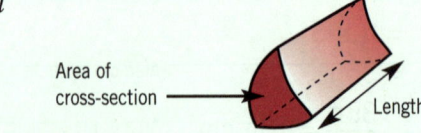

Volume of a cylinder

Cylinders are prisms with circular cross-sections.
Volume = area of cross-section × height (or length)
$V = \pi r^2 \times h$

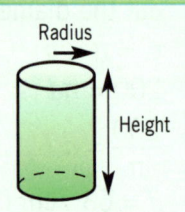

Examples
Find the volume of the following 3D shapes.
Give your answers to 3 significant figures, where appropriate. Use π = 3.142

a) $V = l \times w \times h$
   = 6 × 2 × 3
   = 36cm³

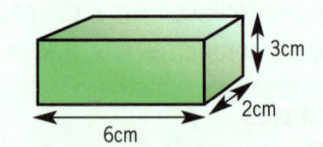

b) $V = a \times l$
   = ($\frac{1}{2}$ × b × h) × l
   = ($\frac{1}{2}$ × 10 × 7) × 15
   = 525cm³

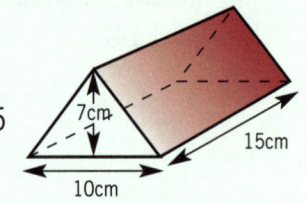

c) $V = \pi r^2 \times h$
   = 3.142 × 10.7² × 24.1
   = 8669.43cm³
   = 8670cm³

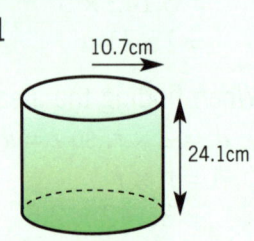

Converting volume units

The cube opposite has a length of 1m – this is the same as a length of 100cm.
Therefore 1m³ = 100 × 100 × 100 = 1 000 000cm³.
Not quite what you may think!

💡 *Converting volume units is a tricky topic that usually catches everybody out. Change all the lengths to the same unit before starting a question.*

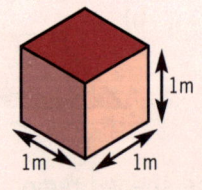

Volume = 1m³

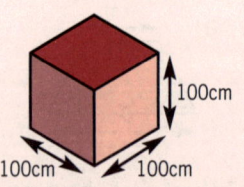

Volume = 1 000 000cm³

72

Surface area of prisms

You need to be able to find the surface area of different prisms.

Cuboid

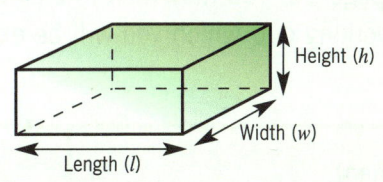

Height (h), Width (w), Length (l)

To find the surface area (SA) of a cuboid, work out the area of each face and then add them together.

$$SA = 2hl + 2hw + 2lw$$

Cylinder

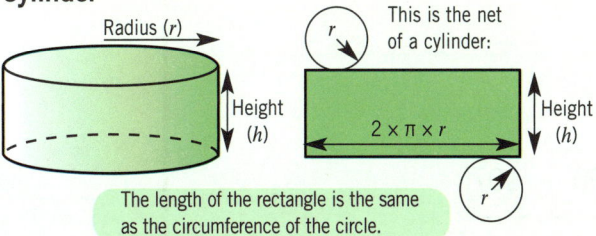

Radius (r), Height (h)

This is the net of a cylinder:

$2 \times \pi \times r$

The length of the rectangle is the same as the circumference of the circle.

$$SA = \underbrace{2\pi rh}_{\text{Area of rectangle}} + \underbrace{2\pi r^2}_{\text{Two circles}}$$

Examples

a) Find the surface area of this cylinder. Use the π button. Give your answer to 1 decimal place.

$SA = 2 \times \pi \times 6 \times 10 + 2 \times \pi \times 6^2$
$= 376.99... + 226.19...$
$= 603.2 \text{cm}^2$

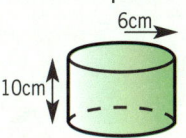

b) A toy tent is shaped as a triangular prism. Work out the amount of material needed to make the tent.

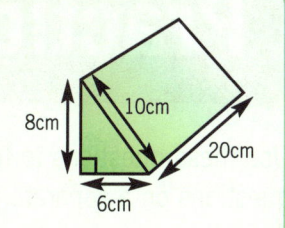

❶ Draw a net of the triangular prism.

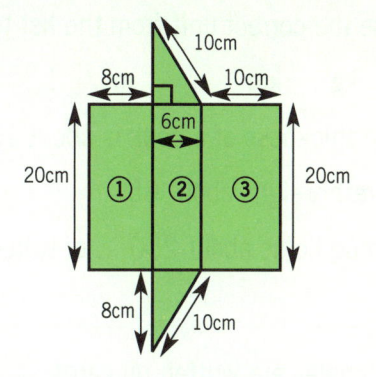

❷ Work out the area of each face.
Triangles (× 2): $A = \frac{1}{2} \times b \times h$
$= \frac{1}{2} \times 6 \times 8$
$= 24\text{cm}^2 \times 2 = 48\text{cm}^2$

Rectangle ①: $A = l \times w$
$= 20 \times 8 = 160 \text{cm}^2$

Rectangle ②: $A = l \times w$
$= 20 \times 6 = 120 \text{cm}^2$

Rectangle ③: $A = l \times w$
$= 20 \times 10 = 200 \text{cm}^2$

❸ Add them together.
Total amount of material needed
$= 48 + 160 + 120 + 200 = 528 \text{cm}^2$

Volumes of enlargements

For an enlargement of **scale factor n**, the volumes are **n^3 times bigger**.

For example, if a cube of length 1cm is enlarged by a scale factor of 2:
$n = 2$, so $V = 2^3 = 8$ times bigger

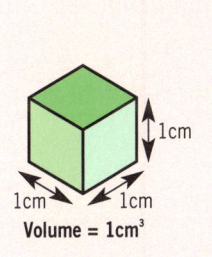

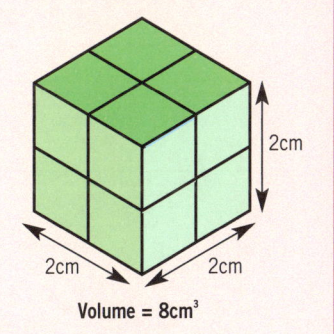

Volume = 1cm³   Volume = 8cm³

Quick test

❶ Work out the volumes of these 3D shapes. Give your answers to 3 significant figures. Use the π key on your calculator.

a) 6.5cm, 27.2cm, 19.8cm

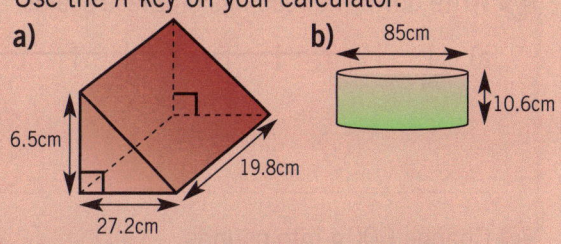

b) 85cm, 10.6cm

❷ Work out the total surface area of the following triangular prism.

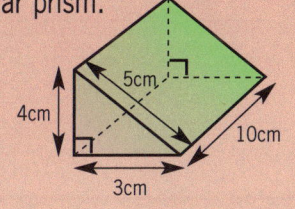

4cm, 5cm, 3cm, 10cm

Geometry and measures

73

Practice questions

Use these questions to test your progress. Check your answers on page 96. You may wish to answer these questions on a separate piece of paper so that you can show full working out, which you will be expected to do in the exam.

1 Choose the correct unit from the list to complete each statement.

cm   kg   km   g   ml   l   m   mm

a) The thickness of a book is about 15 ...

b) Gareth weighs about 65 ...

c) A mug holds about 250 ... of water.

2 Some angles are written on cards:

64°   72°   146°   327°   90°   107°

Which of these angles are...
a) acute?   b) obtuse?   c) reflex?   d) a right angle?

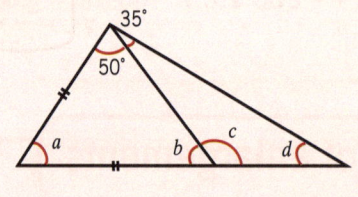

3 A house plan has a scale of 1 : 50. If the width of the house is 42cm on the plan, what is the real width of the house?

4 Calculate the sizes of the angles marked with letters.

a
b
c
d

5 Write down the readings on these scales.

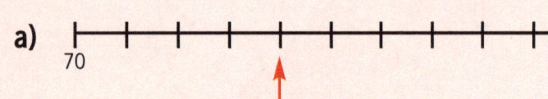

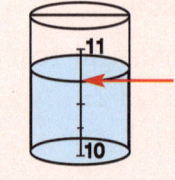

a) (scale from 70 to 80)

b) (cylinder showing reading between 10 and 11)

6 Change 600g into pounds.

7 The scale of a road map is 1 : 50 000. Amersham and Watford are 30cm apart on the map. Work out the real distance in km between Amersham and Watford.

8 Work out the areas of these shapes. Give your answers to 1 d.p.
   a) 9.8m, 12.5m
   b) 8.1cm, 12.4cm, 6.2cm

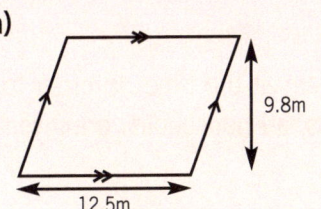

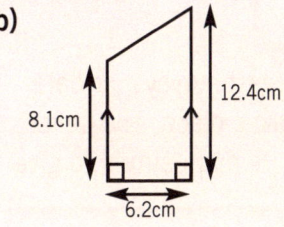

9 A car travels a distance of 320 miles at an average speed of 65mph. How long does it take?

10 A car travels 70 miles in 1 hour 20 minutes. Find the average speed in mph.

11 Jerry said, 'The distance between Manchester and London is 240 miles to the nearest whole mile.' Write down the smallest possible distance between Manchester and London.

12 A ladder of length 6m rests so that the foot of the ladder is 3m away from a wall. Calculate how far up the wall the ladder reaches. Give your answer to 2 s.f.

13 Calculate the volume of the oil drum, clearly stating your units. Use π = 3.142. Give your answer to 3 significant figures.
   125cm, 1.58m

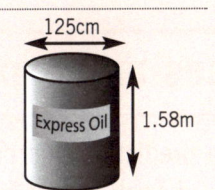

14 James wants to turf his garden. The shape of the garden is shown. Turf costs £5 per square metre. You can only buy a whole number of square metres. How much will the turf cost for James' garden?
   4m, 6m, 5m, 9m

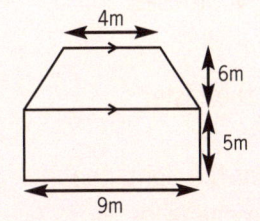

15 The volume of a cuboid is 480cm³. The height of the cuboid is 20cm. The length of the base is 2cm longer than the width of the base. Find the width of the base of the cuboid.

How well did you do?

| 0–4 | Try again | 5–8 | Getting there | 9–12 | Good work | 13–15 | Excellent! |

Collecting data

Data collection

The census is one of the largest surveys, or data collection processes, that takes place. The census is done every 10 years and its main aim is to give a 'snapshot' of Britain at the time. In order to carry out the census, all households are given a survey to complete.

Types of data

There are two main types of data:

Quantitative
The answer is a number, e.g. How many blue cars are there in a car park?

Qualitative
The answer is a word, e.g. What is your favourite colour?

Quantitative data can be discrete or continuous:
- **Discrete data** has an exact value. Each category is separate and is usually found by counting. An example is the number of people with brown hair.

- **Continuous data** has values that merge from one category to the next. Examples include the heights and weights of students. Continuous data cannot be measured exactly. The accuracy of the measurement relies on the accuracy of the measuring equipment.

Primary data is data that is collected by the person who is going to analyse and use it.

Secondary data is data that is available from an external source, such as books, newspapers and the Internet.

Hypotheses and experiments

A **hypothesis** is a **prediction** that can be tested. Experiments can be used to test hypotheses.

For example:
Hypothesis: The better the light, the faster seedlings grow.
Variable: This is the intensity of the light, the condition that can be changed.

Conditions: The other conditions must stay the same. All seedlings must be exactly the same size, strength and colour to start with. If there is **bias** (e.g. if one side of the tray gets more water), then the experiment needs to start again.

Two-way tables

Data can be collected and displayed in a two-way table. This is a table in which data can be read horizontally and vertically.

For example, the table shows the results of a survey of the languages studied by students in a school.

	French	Spanish	German	Total
Female	16	9	1	26
Male	12	12	8	32
Total	28	21	9	58

Questionnaires

Questionnaires can be used to test hypotheses.

When designing questionnaires…
- decide what you need to find out: the **hypothesis**
- give instructions on how the questionnaire has to be filled in
- do not ask for information that is not needed (e.g. name)
- make the questions clear and concise
- keep the questionnaire short
- if people's opinions are needed, make sure the question is **unbiased**. An example of a biased question would be: 'Do you agree that a leisure centre should have a tennis court rather than a squash court?'
- allow for all possible answers. For example, if you were asking 'Which of these is your favourite colour?', add an option for 'other'

Red   Blue   Green   Yellow   Other
☐   ☐   ☐   ☐   ☐

- include a time period if it is needed, for example, 'For how many hours do you play computer games each week?'.

When asked to design a questionnaire, always word the questions carefully. Try to avoid bias appearing in your questions.

Data collection sheet

When collecting data, a data collection sheet is often used. For example, Tracey and David carried out a survey of the colour of cars that passed the gates of their school during a 30-minute interval. Their data collection sheet looked like this:

Colour of car	Tally	Frequency

A tally is a series of lines grouped in fives. The fifth line forms a gate ⅢⅠ.

Siân carried out a survey for the school library. She wanted to find out the type of books students were reading. Her data collection sheet looked like this:

Type of book	Tally	Frequency

Statistics and probability

Quick test

1. Richard and Tammy are carrying out a survey to find out what students' favourite foods are. Design a data collection sheet that they could use.

2. Design a questionnaire you could give to a friend in order to find out what they do in their spare time.

3. Katie wrote this question in her survey:
   'Watching too much television is bad for you. Don't you agree?'   Yes ☐   No ☐
   Explain why this is not a good question.

77

Representing data 1

Pictograms

A **pictogram** can be used to represent discrete data. Pictograms use identical symbols, where each symbol represents a certain number of items.

For example:

Manchester	○ ○ ◐
Luxor	○ ○ ○ ○ ○ ○
Melbourne	○ ◐

Key ○ = 2 hours of sunshine

The pictogram shows the number of hours of sunshine one day in July. From the pictogram, you can see that...
- Melbourne had the fewest hours of sunshine
- Luxor had 12 hours of sunshine
- Manchester had 5 hours of sunshine.

Bar charts

Bar charts have bars of equal width to represent the frequency of discrete data. The width of each bar must be the same.

The bar chart opposite shows the favourite drinks of some students. From the bar chart, you can see that...
- cola is the favourite drink
- milk is the least favourite drink
- three more students prefer water to lemonade
- 8 + 11 + 2 + 5 = 26 students took part in the survey.

Make sure the width of the bars is equal.

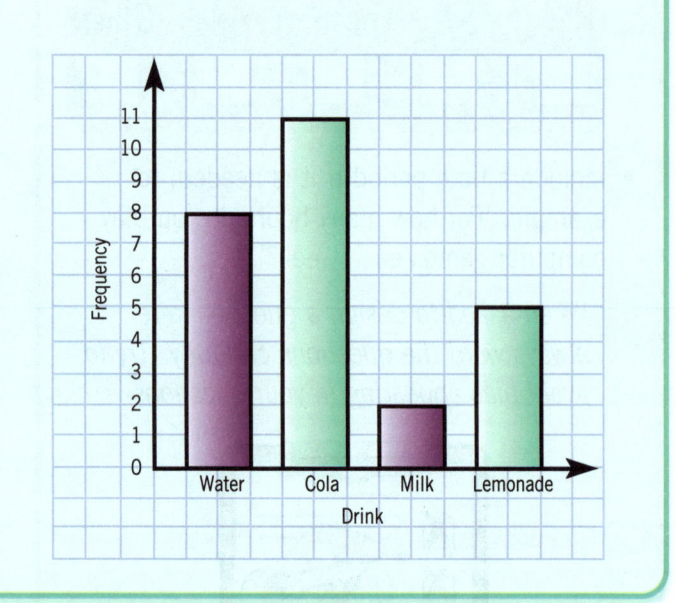

Line graphs

A **line graph** is a set of points joined by lines.

For example, the data below has been used to produce the line graph opposite.

| Year | 2005 | 2006 | 2007 | 2008 | 2009 | 2010 |
| Number of cars sold | 2500 | 2900 | 2100 | 1900 | 1600 | 800 |

Middle values, like point A, have no meaning. Point A does **not** show that halfway between 2007 and 2008, there were 2000 cars sold.

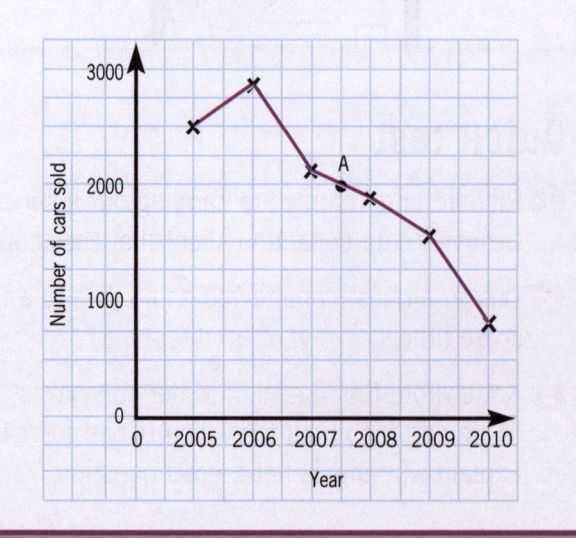

Dual and composite bar charts

Dual and composite bar charts are used to compare data. Dual bar charts have the bars side by side, whereas composite bar charts have the bars on top of each other.

For example, Mr Morris, a PE teacher, wants to find out which sports his students prefer so that he can decide which sport to offer at the after-school club. The results of his survey are shown in the dual bar chart below.

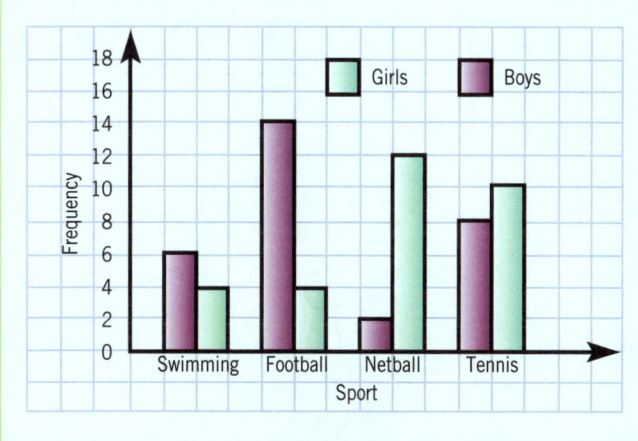

The same results are shown in the composite bar chart below.

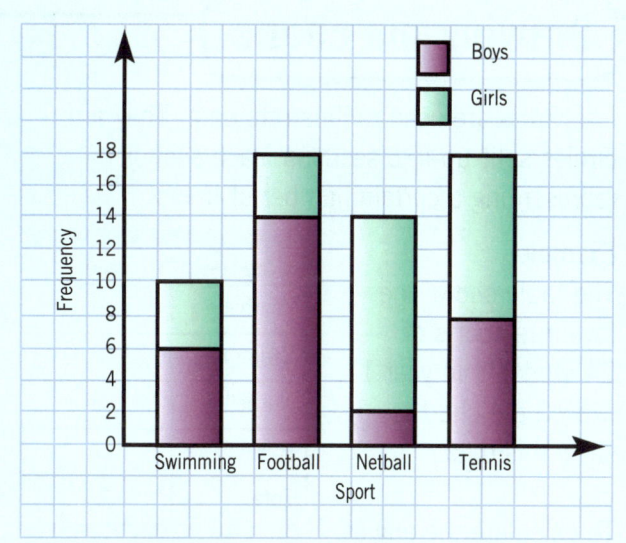

From the bar charts, you can see that...
- if an equal number of boys and girls attend the after-school club, Mr Morris should offer swimming and/or tennis
- if more boys attend the after-school club, Mr Morris should offer football
- if more girls attend the after-school club, Mr Morris should offer netball.

Quick test

1. The incomplete pictogram shows the number of letters sent to a business on four consecutive days.

Monday	✉ ✉ ✉ ✉ ✉
Tuesday	✉ ✉ ⊏
Wednesday	✉ ✉
Thursday	

Key: ✉ = 10 letters

   a) How many letters arrived on Monday?
   b) How many more letters arrived on Tuesday compared with Wednesday?
   c) In total, 125 letters were received between Monday and Thursday. Complete the pictogram.
   d) Explain why you think more letters arrived on the Monday than any other day?

Representing data 2

Drawing pie charts

Pie charts are used to illustrate data. They are circles split up into sections, each section representing a certain number of items.

Example
The table shows the favourite sports of 24 students in year 11. Draw a pie chart to show this data.

Sport	Frequency
Football	9
Swimming	5
Netball	3
Hockey	7
Total	24

To calculate the angles for the pie chart, follow these steps:
1. Find the total for the items listed.
2. Find the fraction of the total for each item.
3. Multiply the fraction by 360° to find the angle.

Sport	Frequency	Angle	Workings
Football	9	135°	$\frac{9}{24} \times 360°$
Swimming	5	75°	$\frac{5}{24} \times 360°$
Netball	3	45°	$\frac{3}{24} \times 360°$
Hockey	7	105°	$\frac{7}{24} \times 360°$
Total	24	360°	

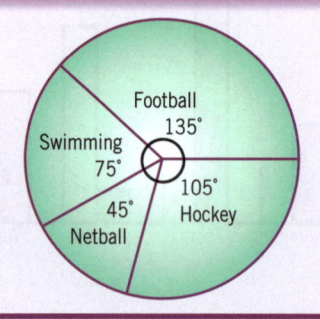

Interpreting pie charts

When the total number of items is known, the number of items in each category can be found.

Example
The pie chart shows how 18 students travel to school.

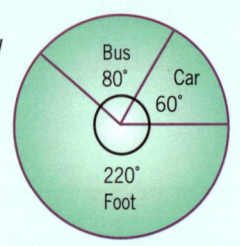

18 students = 360°
1 student = $\frac{360}{18}$
1 student = 20°

How many students travel by...
a) car?
$\frac{60°}{20°}$ = 3 students

b) bus? $\frac{80°}{20°}$ = 4 students

c) foot? $\frac{220°}{20°}$ = 11 students

💡 *Pie chart questions are usually worth about 4 marks at GCSE. Make sure the angles add up to 360° before drawing a pie chart. Measure the angles carefully since you are allowed only a 2° tolerance.*

Comparing pie charts

Pie charts can also be used to compare information.

Example
Jackie is an editor for two magazines. She wants to publish an article on a recently-formed boy band. The pie charts below show the age profile (in years) of the readers of both magazines.

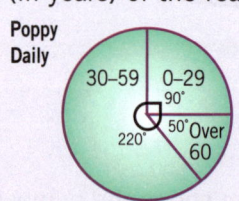

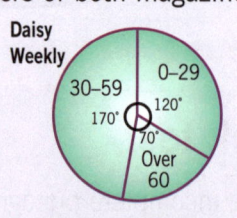

After looking at the pie charts, Jackie decides to insert the article in *Daisy Weekly* because the 0–29 age group is bigger, so there are more younger readers. Explain why Jackie's decision might not be correct.

Jackie might not be correct because the data does not show how many readers there are in total for each magazine. For example, if *Poppy Daily* had 400 readers and *Daisy Weekly* had 90 readers, then 100 people in the 0–29 age group would read *Poppy Daily*, compared with only 30 people in the same age group reading *Daisy Weekly*.

Histograms

Histograms illustrate **continuous data**. They are similar to bar charts except that there are no gaps between the bars. When data is grouped into equal **class intervals** the length (height) of the bars represents the frequency.

For example, the masses of 30 workers in a factory are shown in the table opposite.

Mass (W kg)	Frequency
$45 \leqslant W < 55$	7
$55 \leqslant W < 65$	13
$65 \leqslant W < 75$	6
$75 \leqslant W < 85$	4

$45 \leqslant W < 55$, etc. is called a class interval. In this example the class intervals are all equal in width. $45 \leqslant W < 55$ means the masses are at least 45kg but less than 55kg. A mass of 55kg would be in the next class interval.

Remember these points about histograms:
- The axes do not need to start at zero.
- The axes are labelled.
- The graph has a title.

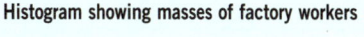

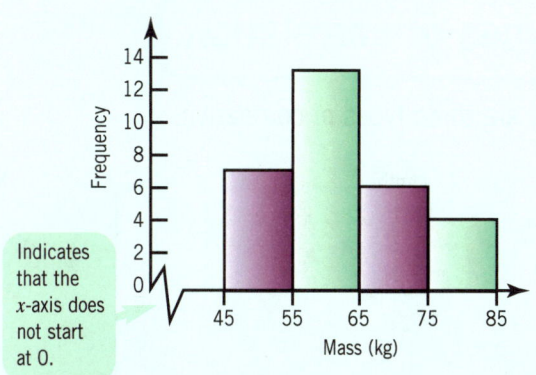

Indicates that the x-axis does not start at 0.

Frequency polygons

To draw a **frequency polygon**, join the **midpoints** of **class intervals** for grouped or continuous data.

Consider the histogram of the factory workers again. To draw a frequency polygon, put a cross at the middle of the top of each bar and join up the crosses with a ruler, as shown opposite. Make sure the frequency polygon is labelled.

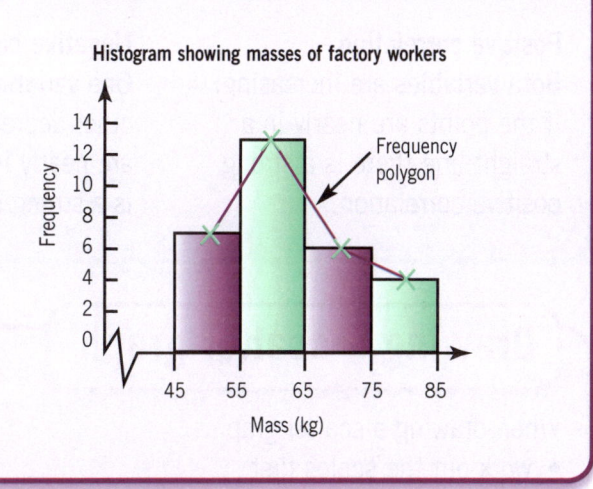

Quick test

Hair colour	Brown	Auburn	Blonde	Black
Frequency	8	4	6	6

1. Draw a pie chart for the data about hair colour opposite.

2. a) Use the histogram to complete the frequency table opposite.
   b) How many students were in the survey?
   c) Draw a frequency polygon on the histogram.

Height (h cm)	Frequency
$140 \leqslant h < 145$	
$145 \leqslant h < 150$	10
$150 \leqslant h < 155$	
$155 \leqslant h < 160$	
$160 \leqslant h < 165$	

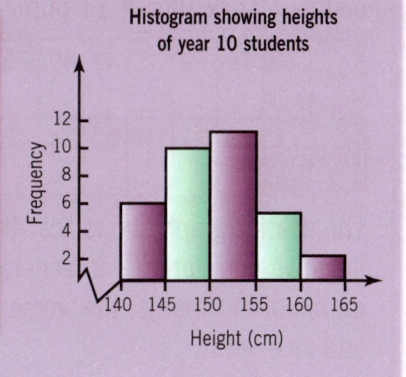

Statistics and probability

81

Scatter graphs & correlation

Scatter graph

A **scatter graph** (scatter diagram or scatter plot) is used to show two sets of data at the same time. Its importance is to show the **correlation** (connection), if any, between two sets of data.

Types of correlation

There are three types of correlation:

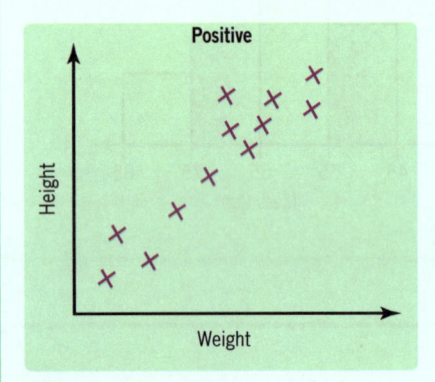

Positive correlation
Both variables are increasing. If the points are nearly in a straight line there is a strong positive correlation.

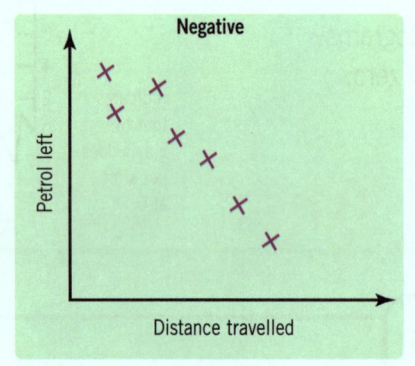

Negative correlation
One variable increases as the other decreases. If the points are nearly in a straight line there is a strong negative correlation.

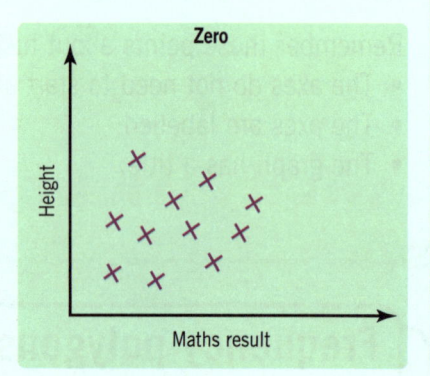

Zero/no correlation
There is little or no linear relationship between the variables.

Drawing a scatter graph

When drawing a scatter graph...
- work out the scales first
- plot the points carefully
- each time a point is plotted, tick it off your list of data.

For example, the table shows the maths and history test results of 11 pupils.

Maths test (%)	76	79	38	42	49	75	83	82	66	61	54
History test (%)	70	36	84	70	74	42	29	33	50	56	64

The scatter graph shows that there is a strong negative correlation – in general, the better the pupils did in maths, the worse they did in history, and vice versa.

Statistics and probability

Lines of best fit

The **line of best fit** is the line that best fits the data points on the scatter graph. It goes in the direction of the data and there is roughly the same number of points above the line as below it. It does not have to pass through any points. A line of best fit can be used to make predictions.

For example, if Hassam was away for the maths test but got 78% in the history test, then from the scatter graph you can estimate that he would have got approximately 43% in maths.

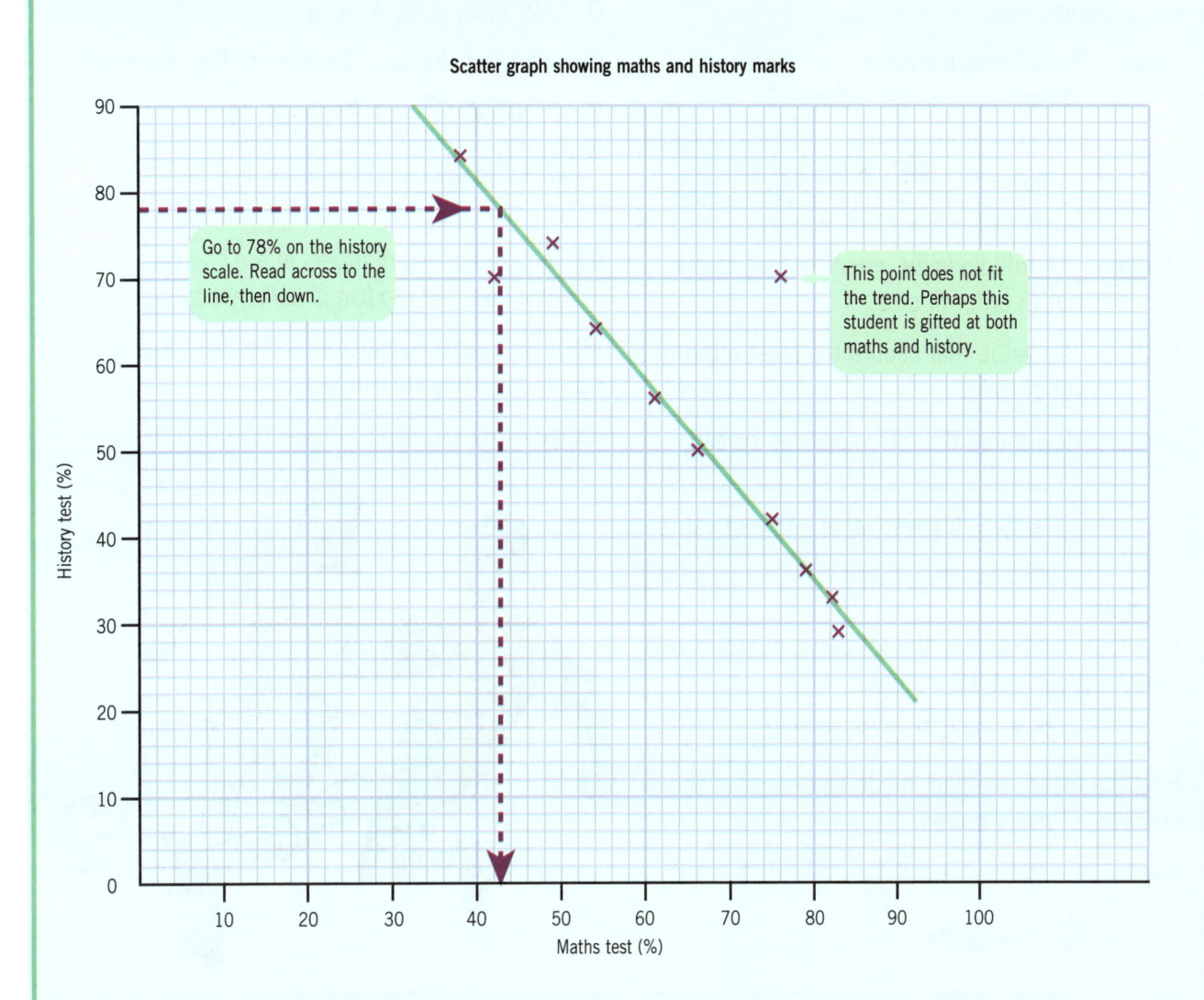

Scatter graph showing maths and history marks

(Go to 78% on the history scale. Read across to the line, then down.)

(This point does not fit the trend. Perhaps this student is gifted at both maths and history.)

> It is not wise to use the line of best fit outside the group of plotted points. For example, a person who got 5% in the history test would not get 106% for maths!

> Do not rush when drawing a scatter graph. Plot the points very carefully. Remember to show on your graph how you made your estimates.

Quick test

1. Write down what type of correlation you would expect for each pair of variables.
    a) The number of pages in a magazine and the number of advertisements.
    b) The heights of students in a year group and their marks in a maths test.
    c) The height up a mountain and the temperature.
    d) The age of a used car and its value.

Averages 1

Averages of discrete data

You should know three types of average: **mean**, **median** and **mode**.

Mean: Sometimes known as the 'average'. The symbol for the mean is $\bar{x}$.

$$\text{Mean} = \frac{\text{sum of a set of values}}{\text{the number of values used}}$$

Median: The **middle value** when the values are put in order of size.

Mode: The value that occurs the **most often** in a set of data.

Range: How much the information is **spread**.

$$\text{Range} = \text{highest value} - \text{lowest value}$$

Example
A football team scored the following numbers of goals in their first 10 matches.

2, 4, 0, 1, 2, 2, 3, 6, 2, 4

Find the mean, median, mode and range of the numbers of goals scored.

Mean = $\dfrac{2+4+0+1+2+2+3+6+2+4}{10}$

= $\dfrac{26}{10}$ = 2.6 goals

Median: put the data in order of size =
0, 1, 2, 2, 2, 2, 3, 4, 4, 6

Cross off in pairs from the ends to find the middle.

0̶ 1̶ 2̶ 2̶ ②　② 3̶ 4̶ 4̶ 6̶

$\dfrac{2+2}{2}$ = 2 goals

If there are two numbers in the middle of a set of values, the median is halfway between them.

Mode = 2 goals because it is the most frequent score, occurring four times.

Range = highest score – lowest score = 6 – 0 = 6

Finding a missing value when given the mean

If you are given the mean of a set of discrete data, you can use the information to calculate a missing value.

Example
A manufacturer produces boxes of matches. He wishes to make the claim, 'Average of 48 matches per box'. In three out of the first four boxes he looks at, the number of matches is 45, 44 and 51. How many matches need to be in the fourth box in order for the manufacturer to make his claim?

Call the number of matches in the fourth box, y.

$\dfrac{45 + 44 + 51 + y}{4} = 48$

$\dfrac{140 + y}{4} = 48$

The mean number of matches must be 48.

$140 + y = 48 \times 4$
$140 + y = 192$
$y = 52$

He needs 52 matches in the fourth box in order to make his claim.

Finding averages from a frequency table

A **frequency table** tells you how many data items there are in a group. For example:

Number of sisters (x)	0	1	2	3	4
Frequency (f)	4	9	3	5	2

This means 2 people had 4 sisters.

Mean: $(\bar{x}) = \frac{\sum fx}{\sum f}$ ($\sum$ means 'the sum of')

$= \frac{(4 \times 0) + (9 \times 1) + (3 \times 2) + (5 \times 3) + (2 \times 4)}{4 + 9 + 3 + 5 + 2}$

$= \frac{38}{23} = 1.7$ (1 d.p.)

💡 When finding the mean from a frequency table, remember to divide by the sum of the frequencies and not by how many groups there are.

Median: Since there are 23 people who have been asked, the median will be the 12th person in the table.

11 people	12th person	11 people

The 12th person has 1 sister – the median = 1

Mode: This is the data value with the highest frequency, that is 1 sister.

Range: 4 – 0 = 4   *Highest number of sisters – lowest number of sisters*

Example

The table gives information on the number of minutes taken by different people to solve a problem.

Number of minutes (x)	0	1	2	3	4	5
Frequency (f)	1	5	6	10	3	1

Find the mean, median, mode and range of the number of minutes taken to solve the problem.

Mean: $(\bar{x}) = \frac{\sum fx}{\sum f}$

$= \frac{(1 \times 0) + (5 \times 1) + (6 \times 2) + (10 \times 3) + (3 \times 4) + (1 \times 5)}{1 + 5 + 6 + 10 + 3 + 1}$

$= \frac{0 + 5 + 12 + 30 + 12 + 5}{26}$

$= \frac{64}{26}$

$= 2.46$ minutes (2 d.p.)

Median: Since there are 26 people who solved the problem, the median is between the 13th and 14th person, who both took 3 minutes.

12 people	13th/14th person	12 people

– the median = 3 minutes

Mode: 3 minutes, as this is the value with the highest frequency.

Range: 5 – 0 = 5 minutes

Quick test

1. Find the mean, median, mode and range of this set of data. 🖩
   2, 9, 3, 6, 4, 4, 5, 8, 4

2. Charlotte made this table to show the number of minutes students were late for registration.

Number of minutes late (x)	0	1	2	3	4
Frequency (f)	10	4	6	3	2

   Calculate... 🖩
   a) the mean   b) the median   c) the mode   d) the range.

Statistics and probability

85

Averages 2

Averages of grouped data

When data values are grouped into class intervals, the exact data values are not known. You can only estimate the mean by using the **midpoint** of the **class interval**. The midpoint is the halfway value.

When you are using grouped (continuous) data, you can only find the **modal class** as opposed to the mode. This is the class interval with the highest frequency.

💡 *Finding the mean of grouped data is a very common GCSE question and is usually worth about 4 marks.*

For example, the table below shows the masses of some year 9 students.

Mass (M kg)	Frequency (f)	Midpoint (x)	fx
40 ≤ M < 45	7	42.5	297.5
45 ≤ M < 50	4	47.5	190
50 ≤ M < 55	3	52.5	157.5
55 ≤ M < 60	1	57.5	57.5

Adding these extra columns helps to show your working out.

Mean = $\frac{\sum fx}{\sum f}$

= $\frac{(7 \times 42.5) + (4 \times 47.5) + (3 \times 52.5) + (1 \times 57.5)}{7 + 4 + 3 + 1}$

= $\frac{702.5}{15}$ = 46.8kg (1 d.p.)

This method is the same as on page 85 except that the frequency is multiplied by the **midpoint of each class interval**.

The **modal class** is 40 ≤ M < 45.

Median
How to find the class interval containing the median:

The middle value of 15 items will be the eighth item. So, the median will be the eighth year 9 student.

Mass (M kg)	Frequency	
40 ≤ M < 45	7	1st to 7th student
45 ≤ M < 50	4	8th to 11th student

The median lies in the class interval 45 ≤ M < 50.

💡 *If your calculator will do statistical calculations, learn how to use these functions. It is much quicker but always do it twice as a check. Always try to show full working out in order to obtain some method marks.*

Using appropriate averages

Choosing which average to use depends on the type of data you have and what you are looking for in it:

- The **mean** is useful when you need a 'typical' value. Be careful not to use the mean if there are extreme values.
- The **median** is a useful average if there are extreme values.
- The **mode** is useful when you need the most common value.

Stem-and-leaf diagrams

Stem-and-leaf diagrams are used for recording and displaying information. They can also be used to find the mode, median and range of a set of data.

For example, these are the marks gained by some students in a mathematics exam:

24	61	55	36	42
32	60	51	38	58
55	52	47	55	55

When the information is put into a stem-and-leaf diagram it looks like this:

Stem	Leaf
2	4
3	2 6 8
4	2 7
5	1 2 5 5 5 5 8
6	0 1

Stem is 30, leaf is 2 – 32

Key: 4 | 2 = 42 marks

To read off the values you multiply the stem by 10 and add on the leaf.

Using the stem-and-leaf diagram, the mode, median and range can be found easily.

Mode = 55
Median = 52
Range = 61 – 24 = 37

Back-to-back stem-and-leaf diagrams are particularly useful when comparing two sets of data. This back-to-back stem-and-leaf diagram shows the times, to the nearest minute, taken by some students to complete a logic problem.

Boys' times		Girls' times
7 7 5 4 0	1	2 5 5 7
6 6 3 1 1	2	3 8 8 9
8 7	3	0 1 2 3 9
9 7 5	4	2 3

Key for boys' times: 7 | 1 = 17 minutes
Key for girls' times: 1 | 5 = 15 minutes

Using averages and spread to compare distributions

Be careful when drawing conclusions from averages, as they do not always tell the whole story.

For example, the lifetime of two types of battery is tested. The mean lifetime of Trojan batteries is 9.2 hours. The mean lifetime of Warrior batteries is 11.6 hours.

From the averages, you might say that all Warrior batteries last longer than Trojan batteries. However, if you look at the range for each type of battery, you can see that this is not true:

Range of Trojan battery = 14.6 hours – 5.8 hours
= 8.8 hours

Range of Warrior battery = 13.2 hours – 9.3 hours
= 3.9 hours

Using the range, you can see that not all Warrior batteries last longer than Trojan batteries. The average lifetime of the Trojan batteries has been reduced because some of them have a particularly low lifetime.

Quick test

1. The heights of some year 10 pupils are shown in the table.

Height (h cm)	$140 \leq h < 145$	$145 \leq h < 150$	$150 \leq h < 155$	$155 \leq h < 160$	$160 \leq h < 165$
Frequency	4	7	14	5	2

a) Calculate an estimate for the mean of this data. Give your answer to 2 decimal places.
b) Write down the modal class.
c) Write down the class interval that contains the median.

Probability 1

What is probability?

Probability is the chance or likelihood that something will happen. All probabilities lie from 0 to 1 inclusive.

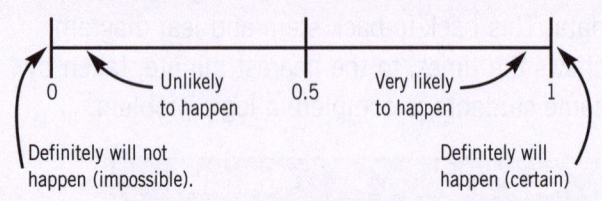

```
  0        Unlikely      0.5      Very likely        1
          to happen              to happen
Definitely will not                        Definitely will
happen (impossible).                       happen (certain)
```

An event with a probability of 0.62 is more likely to happen than an event with a probability of 0.6, since 0.62 is nearer to 1.

Exhaustive events account for all possible outcomes. For example, the list HH, HT, TH, TT gives all possible outcomes when two coins are tossed simultaneously.

Mutually exclusive events or outcomes cannot happen at the same time. For example, if a student is chosen at random:

Event A: the student is a girl
Event B: the student is a boy
These are **mutually exclusive** because no student can be both a boy and a girl.

Event C: the student has brown hair
Event D: the student wears glasses
These are **not mutually exclusive** because brown-haired students may wear glasses.

$$\text{Probability of an event} = \frac{\text{number of times the event can happen}}{\text{total number of possible outcomes}}$$

P(event) is a shortened way of writing the probability of an event.

Example
There are six blue, four yellow and two red beads in a bag. John chooses a bead at random. What is the probability he chooses...

a) a red bead?

$P(\text{red}) = \frac{2}{12}$ or $\frac{1}{6}$

b) a yellow bead?

$P(\text{yellow}) = \frac{4}{12}$ or $\frac{1}{3}$

c) a blue, yellow or red bead?

$P(\text{blue, yellow or red}) = \frac{12}{12} = 1$

d) a white bead?

$P(\text{white}) = 0$

💡 *Probabilities must be written as a fraction, decimal or percentage. Probabilities can never be negative or greater than 1.*

Expected outcome

The expected outcome of an event is also known as theoretical probability.

Examples
a) A fair dice is thrown 300 times. Approximately how many fives are likely to be obtained?

There are six possible outcomes and all are equally likely.

$P(5) = \frac{1}{6} \times 300 = 50$ fives

Multiply 300 by $\frac{1}{6}$ since a 5 is expected $\frac{1}{6}$ of the time.

b) The probability of passing a driving test at the first attempt is 0.65. If there are 200 people taking their test for the first time, how many would you expect to pass?

$0.65 \times 200 =$ 130 people would be expected to pass

Relative frequencies

Relative frequencies are used to estimate probability. If it is not possible to calculate a probability, an experiment can be used to find the relative frequency.

Relative frequency of an event = $\dfrac{\text{number of times event occurred}}{\text{total number of trials}}$

Example
When a fair dice was thrown 80 times, a six came up 12 times. What is the relative frequency of getting a six?

Number of trials = 80
Number of sixes = 12
Relative frequency = $\dfrac{12}{80}$ = 0.15

Using experiments to calculate relative frequencies

If a fair dice was thrown 180 times it would be expected that 30 twos would be thrown.

$\dfrac{1}{6} \times 180 = 30$

Example
Throw a dice 180 times and record the frequency of twos in every 30 throws.

Number of throws	Total frequency of twos	Relative frequency
30	3	0.1
60	7	0.12
90	16	0.18
120	19	0.16
150	24	0.16
180	31	0.17

It is expected that one-sixth ($\dfrac{1}{6} = 0.1\dot{6} = 0.17$) of the throws will be twos.

As the number of throws increases, the relative frequency gets closer to the expected probability.

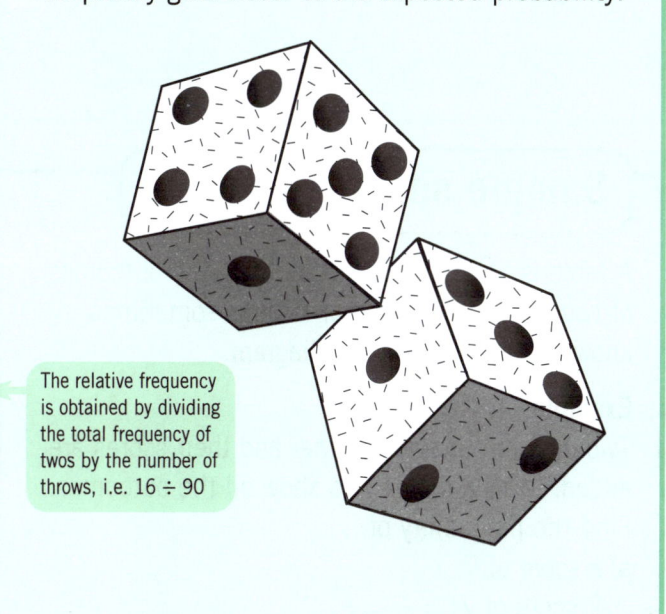

The relative frequency is obtained by dividing the total frequency of twos by the number of throws, i.e. 16 ÷ 90

Quick test

1. Write down an event that will have a probability of zero.

2. A box contains three salt and vinegar, four cheese and two bacon-flavoured packets of crisps. If a packet of crisps is chosen at random, what is the probability that it is...
   a) salt and vinegar?   b) cheese?   c) onion flavoured?

3. The probability of achieving a grade C in mathematics is 0.48 If 500 students sit the exam, how many would you expect to achieve a grade C?

4. When a fair dice was thrown 200 times, a five came up 47 times. What is the relative frequency of getting a five?

Probability 2

Possible outcomes of two or more events

You can use lists, tables and **sample space diagrams** when answering probability questions with two **successive events**.

Lists

Making a systematic list of possible outcomes of two events is useful.

Example
A coin is tossed twice. Make a list of the possible outcomes. What is the probability of getting two heads?

There are four different outcomes:
| HH | TT | HT | TH |
| Both heads | Both tails | 1st head, 2nd tail | 1st tail, 2nd head |

$P(HH) = \frac{1}{4}$

💡 *When there are two events, you need to show all the outcomes clearly – this will make calculating probabilities easier.*

Sample space diagrams

A table is helpful when considering the outcomes of two events. This kind of table is sometimes known as a **sample space diagram**.

Example
Two dice are thrown together and their scores are added. Draw a diagram to show all the outcomes. Find the probability of...

a) a score of 7
   $P(\text{score of } 7) = \frac{6}{36} = \frac{1}{6}$
b) a score that is a multiple of 4
   $P(\text{multiple of } 4) = \frac{9}{36} = \frac{1}{4}$
c) a score of 2 or 3
   $P(\text{score of 2 or 3}) = \frac{3}{36} = \frac{1}{12}$

		First dice					
		1	2	3	4	5	6
Second dice	1	2	3	4	5	6	7
	2	3	4	5	6	7	8
	3	4	5	6	7	8	9
	4	5	6	7	8	9	10
	5	6	7	8	9	10	11
	6	7	8	9	10	11	12

There are 36 possible outcomes.

Two-way tables

Example
The **two-way table** shows the number of students in a class who are left-handed or right-handed.

Hand	Male	Female	Total
Right	14	10	24
Left	2	7	9
Total	16	17	33

a) What is the probability that a person chosen at random is right-handed?

   $P(\text{right-handed}) = \frac{24}{33} = \frac{8}{11}$

b) If a boy is chosen at random, what is the probability that he is left-handed?

   $P(\text{left-handed boy}) = \frac{2}{16} = \frac{1}{8}$

Probability of an event not happening

If two events are **mutually exclusive**, then:
P(event will happen) = 1 − P(event will not happen)
or
P(event will not happen) = 1 − P(event will happen)

Examples

a) The probability that someone will get flu next winter is 0.42. What is the probability that they will not get flu next winter?

P(not get flu) = 1 − P(get flu)
   = 1 − 0.42
   = 0.58

b) Mrs Smith chooses one book from her library each week. She chooses a horror story, a thriller or a crime novel. The probability that she chooses a horror story is 0.35. The probability that she chooses a thriller is 0.25. Work out the probability that Mrs Smith chooses a crime novel.

Probability she chooses a horror story or thriller is 0.35 + 0.25 = 0.6

Probability she chooses a crime novel is
1 − P(does not choose a crime novel)
   = 1 − 0.6
   = 0.4

The multiplication law

When two events are independent the outcome of the second event is not affected by the outcome of the first.

If two or more events are independent the probability of A and B and C… happening together is found by multiplying the separate probabilities.

P(A and B and C…) = P(A) × P(B) × P(C)…

Example
The probability it will rain on any day in August is $\frac{3}{10}$. Find the probability that it will rain on both 1 August and 29 August?

P(rain and rain) = $\frac{3}{10} \times \frac{3}{10} = \frac{9}{100}$

The addition law

If two or more events are **mutually exclusive** the probability of A or B or C… happening is found by **adding** the probabilities.

P(A or B or C…) = P(A) + P(B) + P(C) + …

Example
There are 11 counters in a bag: 5 of the counters are red and 3 of them are white. Lucy picks a counter at random. Find the probability that Lucy's counter is either red or white.

P(red) = $\frac{5}{11}$ P(white) = $\frac{3}{11}$

P(red or white) = P(red) + P(white)

$\frac{5}{11} + \frac{3}{11} = \frac{8}{11}$ Red and white are mutually exclusive

Quick test

1. The probability that it will not rain tomorrow is $\frac{2}{9}$. What is the probability that it will rain tomorrow?
2. a) Draw a sample space diagram that shows the possible outcomes when two dice are thrown together and their scores are multiplied.
   b) What is the probability of getting a score of 6?
   c) What is the probability of getting a score of 37?

Practice questions

Use these questions to test your progress. Check your answers on page 96. You may wish to answer these questons on a separate piece of paper so that you can show full working out, which you will be expected to do in the exam.

1 This pictogram has no key. The pictogram shows the number of cups of tea Mrs Bassett sold in her tea shop during one week. She sold 25 cups of tea on Wednesday.

Sunday
Monday ○ ○ ○ ◖
Tuesday ○ ○
Wednesday ○ ○ ◖
Thursday ○ ○ ◖
Friday ○ ○ ○ ○ ◖
Saturday ○

a) How many cups of tea did Mrs Bassett sell during the week?

b) William says, 'Nobody wanted a cup of tea on Sunday because the weather was too hot.' Give another reason why no cups of tea were sold on Sunday.

2 Reece carried out a survey to find out the favourite flavours of crisps of students in his year. The results are shown in the table (right). Draw a pie chart of this information.

Crisp flavour	Number of students
Cheese and onion	14
Salt and vinegar	20
Beef	12
Smoky bacon	2

3 Find the mean, median and mode of these quantities: 6, 2, 1, 4, 2, 2, 5, 3

4 A bag contains three red, four blue and six green balls. If a ball is chosen at random from the bag, what is the probability of choosing...

a) a red ball?   **b)** a green ball?   **c)** a yellow ball?   **d)** a blue or a red ball?

5 The probability of passing a driving test is 0.7
If 200 people take the test today, how many would you expect to pass?

6 In a survey the heights of ten girls and their shoe sizes were measured:

Height in cm	150	157	159	161	158	164	154	152	162	168
Shoe size	3	5	$5\frac{1}{2}$	6	5	$6\frac{1}{2}$	4	$3\frac{1}{2}$	6	7

a) Draw a scatter graph to illustrate this data.

b) What type of correlation is there between height and shoe size?

c) Draw a line of best fit on your diagram.

d) From your scatter graph, estimate the height of a girl whose shoe size is $4\frac{1}{2}$.

7 Two spinners are used in a game. The first spinner is labelled 2, 4, 6, 8.
The second spinner is labelled 3, 5, 5, 7. Both spinners are spun.
The score is found by multiplying the numbers on each spinner.

a) Complete the table to show the possible scores.

b) What is the probability of getting an even score?

c) What is the probability of getting a score of 10?

	First spinner			
Second spinner	2	4	6	8
3				
5				
5				
7				

8 The masses of some students in a class are measured. The results are given in the table.

a) Work out an estimate for the mean mass of the students.

b) What is the modal class?

Mass (M) in kg	No. of students
$40 \leq M < 45$	6
$45 \leq M < 50$	5
$50 \leq M < 55$	8
$55 \leq M < 60$	4
$60 \leq M < 65$	2

9 Mr Alex has to choose someone to enter a spelling contest. The last 10 scores (out of 10) for Lucy and Rashna are given below:

Lucy: 5, 6, 8, 4, 1, 1, 3, 10, 2, 10
Rashna: 6, 6, 5, 5, 4, 4, 5, 5, 6, 4

a) Work out the mean scores and ranges for Lucy and Rashna.

b) Who should Mr Alex choose to be in the spelling contest and why?

10 Jonathan is doing a survey into whether people play sport. He decides to ask 50 people by standing outside a sports centre on a Wednesday morning. Explain why his survey will be biased.

How well did you do?

| 0–3 | Try again | 4–6 | Getting there | 7–8 | Good work | 9–10 | Excellent! |

Statistics and probability

Answers

Number   Quick test answers

Page 5   Types of numbers
1. 11, 13, 17, 19, 23, 29   2. HCF: 12; LCM: 120
3. a) ±8   b) 6   4. a) $\frac{4}{3}$ b) $\frac{p}{x}$ c) $\frac{1}{5}$ d) 10

Page 7   Positive & negative numbers
1. 3°C   2. a) 4   b) -16   c) -12   d) -12   e) 5   f) 6   g) 7   h) -10   i) 36

Page 9   Working with numbers
1. a) 1177   b) 232   c) 1722   d) 275
2. 20 496   3. 37
4. a) 164   b) 890   c) 37 000   d) 0.097
5. 42 people

Page 11   Fractions
1. a) $\frac{1}{3}$ b) $\frac{7}{20}$ c) $\frac{6}{13}$ d) $1\frac{1}{3}$ e) $\frac{5}{21}$ f) $\frac{4}{21}$ g) $\frac{49}{121}$ h) $\frac{50}{63}$
2. £40   3. 247mm   4. $\frac{4}{15}$

Page 13   Decimals
1. a) 36.48   b) 17.679   c) 26.88   d) 12.3   e) 6   f) 52   g) 0.4   h) 0.0037   i) 4000   j) 45 000   k) 470 000   l) 32 500   m) 693   n) 291   o) 7.074   p) 0.0284
2. a) 2.607, 2.61, 2.615, 4.02, 4.021, 4.20
   b) 6.05, 6.49, 8.206, 8.27, 8.271, 8.93

Page 15   Rounding
1. a) 6.43   b) 18.61   c) 14.27   d) 29.64
2. 6700km

Page 17   Percentages 1
1. £210   2. 74.6%   3. 9.9lb   4. Super's; £33.33

Page 19   Percentages 2
1. £161.50   2. a) £1800   b) £6180   3. £204 120   4. £3394.88

Page 20   Fractions, decimals & percentages
1. a) i) $0.\dot{2}8571\dot{4}$ ii) 0.6   iii) $0.\dot{8}$
   b) i) 28.57% (2 d.p.)   ii) 60%   iii) $88.\dot{8}$%
2. 0.041, 5%, 26%, $\frac{1}{3}$, $\frac{2}{5}$, 0.42

Page 21   Using a calculator
1. a) 14.45   b) 769.6   c) 7.052   d) $8\frac{1}{3}$ or $8.\dot{3}$

Page 23   Approximations & checking calculations
1. 100   2. 10 rolls   3. £7.45 (check: £175 ÷ 25 = £7✓)

Page 25   Ratio
1. a) 4 : 5   b) 1 : 2   c) 5 : 2
2. 10, 15 and 35 sweets respectively   3. £2.76   4. 36cm

Page 27   Indices
1. a) 12^{12} b) 9^6 c) 4   d) 18^4 e) 4^9 f) 1
2. a) x^{13} b) $6x^{16}$ c) $4x^2$ d) $5x^7$ e) $2x^{12}$ f) $16x^{10}$

Pages 28–29   Answers to practice questions
1. a) 9, 21, 41   b) 9, 64, 100   c) 2, 41   d) 2, 40   e) 40, 64, 100
2. 9°C   3. £27   4. a) 56   b) 508   c) 7002
5. 27, 405, 639, 728, 736, 829   6. 77.3%   7. £315
8. 5.032, 5.04, 5.42, 6.27, 6.385, 6.39   9. 4 : 5
10. Bowling – John's taxis: (3.20 × 18) × 2 + (4.45 × 2)
    = 57.60 × 2 + 8.90 = £124.10
    Robert's cars: (2.50 × 18) × 2 + (15 × 2) + (4.45 × 2)
    = 45 × 2 + 30 + 8.90 = £128.90
    Cinema – John's taxis: (3.20 × 13) × 2 + (7.50 × 2)
    = 41.60 × 2 + 15 = £98.20
    Robert's cars: (2.50 × 13) × 2 + (15 × 2) + (7.50 × 2)
    = 32.5 × 2 + 30 + 15 = £110
    Concert – 49.50 × 2 = £99
    Going to the cinema using John's taxis is the cheapest.
11. The 100ml tube of toothpaste. Since 50ml costs 2.48p per ml, 75ml costs 2.61p per ml and 100ml costs 2.42p per ml.
12. 1.5cm   13. £6502.50   14. a) $2^5 = 32$ b) a^7 c) b^4 d) $4a^7$
15. £440.63

Algebra   Quick test answers

Page 31   Algebra
1. a) $10a$ b) $8a + b$ c) $9x + 4y$ d) $4x^2y - 5xy^2$
2. a) $3x + 6$ b) $2x + 2y$ c) $-6x - 12$ d) $8x + 13$ e) $7x + 8$
   f) $10x + 12$
3. a) $2(2x^2 + 1)$ b) $2(y - 2)$ c) $2(3x + 5)$ d) $3x(x + 3)$
   e) $5x(3 - 4x)$ f) $x(x - 5)$

Page 33   Equations 1
1. a) $x = 4$ b) $x = 6$ c) $x = 20$ d) $x = 13$ e) $x = 24$ f) $x = 4.5$
   g) $x = 2$ h) $x = 4$ i) $x = -1$ j) $x = -1.4$

Page 35   Equations 2
1. $x = 3, y = -1$
2. 3.3
3. a) $4(x + 1) = 4x + 4$ b) $x = 9$ – length = 11cm

Page 37   Patterns, sequences & inequalities
1. a) 13, 15   b) 25, 36   c) 4, 2
2. a) $2n + 3$ b) $3n - 1$ c) $4n + 2$ d) $10 - 2n$
3. a) $x < 6$ b) $x \geq 4$ c) $1 \leq x \leq 3$ d) $\frac{-1}{5} \leq x < 2$

Page 39   Formulae
1. a) 10   b) 14   c) 8   d) 6   e) -26
2. $b = \frac{a+d}{5}$ 3. $m = \frac{y-c}{x}$ 4. $R = \frac{V}{I}$

Page 41   Straight-line graphs
1. a) $x = 1$ b) $x = 1.5$
2. $y = 4x - 1$

Page 43   Curved graphs
1. a)

x	-3	-2	-1	0	1	2	3
y	12	7	4	3	4	7	12

b) [graph of $y = x^2 + 3$]

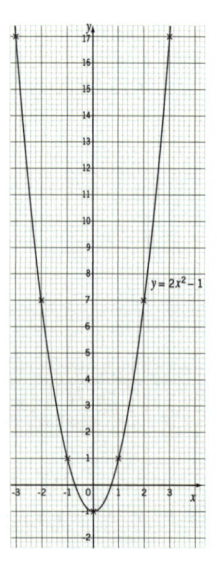

c) $x = -2.2$ and 2.2 approximately

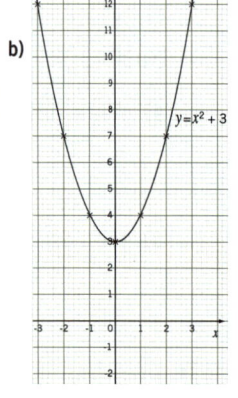

2.

x	-3	-2	-1	0	1	2	3
y	17	7	1	-1	1	7	17

[graph of $y = 2x^2 - 1$]

Page 45 Interpreting graphs
1. Graph 1 = C   Graph 2 = A   Graph 3 = B
2. a) 66.6̇mph   b) 1 hour   c) 50mph   d) About 1442

Pages 46–47 Answers to practice questions
1. $T = 85y + 8z$
2. a) [matchstick pattern diagram]   b)

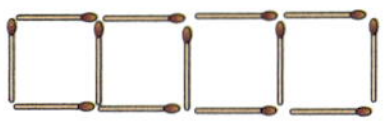

Pattern number (p)	Number of matchsticks (m)
1	4
2	7
3	10
4	13
5	16
6	19

   c) 301 matchsticks   d) $m = 3p + 1$
3. a) $5n + 10$ b) $3n - 6$ c) $2n - 10$ d) $-3n + 18$ e) $8n - 22$
4. a) 21, 25   b) $4n + 1$
5. a) $x = 3$ b) $x = 4$ c) $x = 5$ d) $x = 2$ e) $x = 1$
6. a) $9x + 4 = 22$ b) 3cm
7. a)

x	-2	-1	0	1	2	3
$y = 3x - 4$	-10	-7	-4	-1	2	5

   b) [graph of $y = 3x - 4$]   c) $\left(\frac{4}{3}, 0\right)$

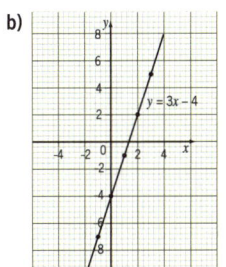

8. £195   9. a) $4a$ b) $5b$ c) $3b$ d) $6c$ e) $8a - b$
10. a) $5(a - 2)$ b) $6(b + 2)$ c) $4y(2y + 3)$
11. a) x^7 b) 6^3 c) $12x^5$ d) x^{12}
12. a) 3   b) 10
13. $n = \frac{m - p}{3}$
14. $x = 1.8$ (1 d.p.)
15. $n \leq 3$
16. Graph 1 = C   Graph 2 = A   Graph 3 = B

Geometry and measures   Quick test answers
Page 49 Shapes
1. Hexagon   2. See information on pages 48–49.

Page 51 Solids
1. [net diagram with 4cm and 5.7cm measurements]
2. [plan, front elevation, side elevation drawings]
   Plan   View from A (front elevation)   Side elevation B

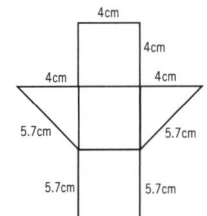

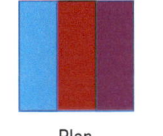

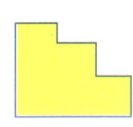

Page 53 Constructions
1. [angle construction 30°/30°]   2. [perpendicular bisector construction, 10 cm]

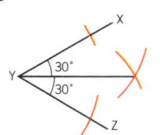

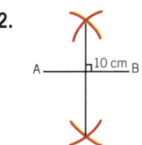

Page 54 Symmetry
1. Line, plane, rotation   2. [symmetry diagram]   3. [two house diagrams] or

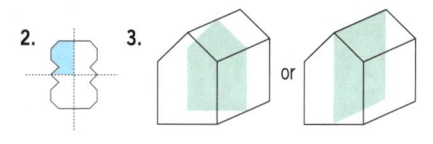

Page 55 Loci & coordinates in 3D
1. [diagram with circle, rectangle RSTU, X marked with 45° angles, 4m radius]

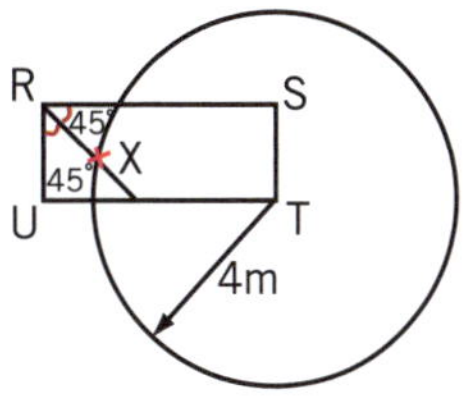

Page 57 Angles
1. a) $a = 150°$ b) $b = 70°, c = 110°, d = 70°$
   c) $a = 50°, b = 50°, c = 130°, d = 50°$
2. a) 72°   b) 108°

Page 59 Bearings & scale drawings
1. a) 072°   b) 305°   c) 145°
2. a) 252°   b) 125°   c) 325°
3. 7km

Page 61 Transformations 1
1. a)–d) [graph showing triangles P, A/B, C, Q, S, R with lines $y = x$ and $y = -1$]

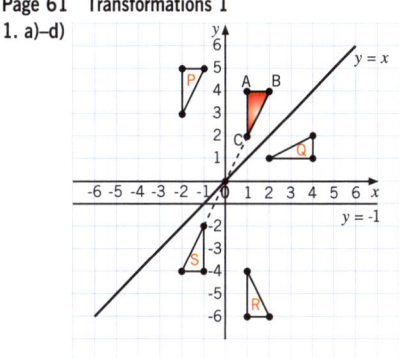

2. Move 2 to the left, 3 up   3. $\binom{-5}{11}$

Page 63 Transformations 2
1. [shape P1 diagram]   2. a)–b) [graph with shapes B and C]

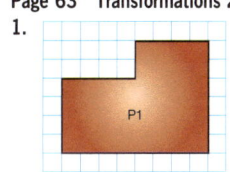

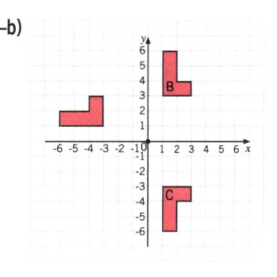

Page 65 Measures & measurement 1
1. 3.5kg   2. 6.6lb   3. 14 pints
4. a) 9.2   b) 9.5   c) 2.42   d) 2.46   e) 2.48   f) 6.25   g) 6.75
5. C (3kg)

Page 67 Measures & measurement 2
1. 9.15secs $\leq t <$ 9.25secs
2. 57.5kg $\leq W <$ 58.5kg
3. 2.25mph
4. 8.5̇7̇ hours (or 8 hours 34 minutes)
5. 2.2g/cm³

Page 69 Pythagoras' theorem
1. a) 17.2cm (1 d.p.)   b) 20.0cm (1 d.p.)   2. 17cm   3. 94.3km (1 d.p.)

Page 71 Area of 2D shapes
1. a) 64cm²   b) 60cm²   c) 63.6cm²   d) 208cm²
2. 21.5cm² (3 s.f.)

Page 73 Volume of 3D shapes
1. a) 1750cm³   b) 60 100cm³
2. 132cm²

Answers

Pages 74–75 Answers to practice questions
1. a) mm b) kg c) ml
2. a) 64°, 72° b) 146°, 107° c) 327° d) 90° 3. 21m
4. $a = 80°$, $b = 50°$, $c = 130°$, $d = 15°$
5. a) 74 b) 10.75 6. 1.32lbs 7. 15km
8. a) 122.5m² b) 63.6cm²
9. 4 hr 55 min 10. 52.5 mph 11. 239.5 miles 12. 5.2m
13. 1 940 000cm³ (= 1940 litres = 1.94m³) 14. £420 15. 4cm

Statistics and probability Quick test answers

Page 77 Collecting data
1.

Favourite food type	Tally	Frequency
List at least four		

2. Make sure your questionnaire follows the guidance given on page 77.
3. This is not a good question because it is a leading question. Katie's opinion is evident and it encourages people to give a particular answer.

Page 79 Representing data 1
1. a) 50 letters
   b) 5 letters
   c) Thursday ✉✉✉
   d) Since there is no post on a Sunday and limited post on a Saturday, there can be a build-up of mail.

Page 81 Representing data 2
1. The angles for the pie chart are: Brown 120°, Auburn 60°, Blonde 90°, Black 90°
2. a) The frequencies for the heights are: $140 \leq h < 145 = 6$, $145 \leq h < 150 = 10$, $150 \leq h < 155 = 11$, $155 \leq h < 160 = 5$, $160 \leq h < 165 = 2$
   b) 34 students
   c) The frequency polygon should be plotted at the midpoints of the bars.

Page 83 Scatter graphs & correlation
1. a) Positive b) Zero/no correlation c) Negative d) Negative

Page 85 Averages 1
1. Mean = 5; median = 4; mode = 4; range = 7
2. a) 1.32 mins b) 1 min c) 0 mins d) 4 mins

Page 87 Averages 2
1. a) 151.56cm b) $150 \leq h < 155$ c) $150 \leq h < 155$

Page 89 Probability 1
1. Any sensible answer, e.g.: I will get a 7 when I throw a dice; I will get a 4 when I toss a coin.
2. a) $\frac{3}{9} = \frac{1}{3}$ b) $\frac{4}{9}$ c) 0
3. 240 students
4. $\frac{47}{200}$ (= 0.235)

Page 91 Probability 2
1. $\frac{7}{9}$
2. a)

	First dice					
		1	2	3	4	5	6
	1	1	2	3	4	5	6
Second	2	2	4	6	8	10	12
dice	3	3	6	9	12	15	18
	4	4	8	12	16	20	24
	5	5	10	15	20	25	30
	6	6	12	18	24	30	36

b) $\frac{4}{36} = \frac{1}{9}$ c) 0

Pages 92–93 Answers to practice questions
1. a) 160 cups of tea b) Shop/cafe closed
2.

Crisp flavour	Number of students	Angle
Cheese and onion	14	105°
Salt and vinegar	20	150°
Beef	12	90°
Smoky bacon	2	15°

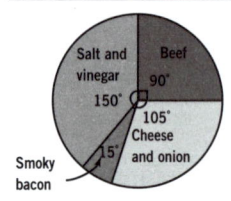

3. Mean = 3.125, Median = 2.5, Mode = 2
4. a) $\frac{3}{13}$ b) $\frac{6}{13}$ c) 0 d) $\frac{7}{13}$
5. 140 people
6. a)

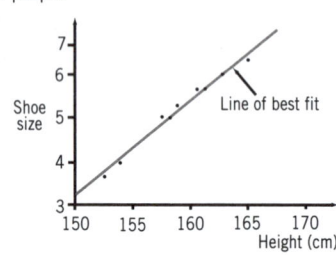

b) Positive correlation c) See scatter graph d) 156cm (allow 155–160)
7. a)

		First spinner			
		2	4	6	8
Second spinner	3	6	12	18	24
	5	10	20	30	40
	5	10	20	30	40
	7	14	28	42	56

b) $\frac{16}{16} = 1$ c) $\frac{2}{16} = \frac{1}{8}$
8. a) 50.7 b) $50 \leq M < 55$
9. Lucy Mean: 5 Rashna Mean: 5
   Range: 9 Range: 2
   Mr Alex should choose Rashna. The mean scores are the same but Rashna is much more consistant because she has the smallest range.
10. People who like sport generally use a sports centre. People who work will not use the sports centre on Wednesday morning.